A Passion for Physics
The Story of a Woman Physicist

A Passion for Physics

The Story of a Woman Physicist

Joan Freeman

Adam Hilger
Bristol, Philadelphia and New York

British Library Cataloguing in Publication Data
Freeman, Joan
   A passion for physics.
   1. Women physicists—Biographies
   I. Title
   530.092

   ISBN 0-7503-0098-1

Library of Congress Cataloging-in-Publication Data
Freeman, Joan, 1918-
   A passion for physics: the story of a woman physicist/Joan Freeman.
      p.   cm.
   Includes index.
   ISBN 0-7503-0098-1 (hbk.)
   1. Freeman, Joan, 1918- . 2. Women scientists. 3. Physicists–
Australia–Biography. I. Title.
QC16.F713A3 1991
530'.092–dc20
[B]
                                                          90-19560
                                                          CIP

Published under the Adam Hilger imprint by IOP Publishing Ltd
Techno House, Redcliffe Way, Bristol BS1 6NX, England
335 East 45th Street, New York, NY 10017-3483, USA

US Editorial Office: 1411 Walnut Street, Philadelphia, PA 19102

Typeset by Keytec Typesetting Ltd, Bridport, Dorset
Printed in Great Britain by Galliard (Printers) Ltd, Norfolk

To John

and to the memory of my mother

Contents

Preface

'You should write your autobiography,' said Lorna Arnold to me one day, some six or so years ago, after I had recounted to her some of my early experiences and struggles on the way to becoming a physicist. Lorna has worked for many years with Professor Margaret Gowing, formerly Professor of the History of Science in the University of Oxford, on the official history (commissioned by the Atomic Energy Authority) of Britain's atomic energy project. I was fortunate to become acquainted with the two historians while I was writing some papers on Harwell history. They have both been an invaluable source of inspiration and encouragement to me, as well as having critically read parts (in Lorna Arnold's case all) of my book in draft.

The idea of writing my own story had not occurred to me until Lorna Arnold suggested it; but then it grew on me. There are still remarkably few women physicists—in the English-speaking world at least. This is the story of one of them, from an early period, and embracing experiences in three continents. It also describes other women physicists I have known, and their individual difficulties and achievements. I have kept scientific details to a minimum hoping that my account will appeal to the general reader as well as to scientists.

The writing of this book has been an absorbing occupation as I revived memories of many exciting moments and interesting people, and reflected on the fascination that the physical world has always held for me. But I know that memory can be unreliable, and I apologise for errors and misrepresentations that must inevitably have crept into my account. I am grateful for the help and encouragement, in conversations and letters, that I have received from many friends and colleagues, too numerous to be named here, besides the following who have read and made suggestions on various drafts and/or procured material and photographs for me: Mrs Nessy Allen, Professor C A Barnes, Professor R J Blin-Stoyle, Miss Diana Bowman, Mrs Hanni Bretscher, Professor W E Burcham, Dr Ursula Bygott, Mr B F C Cooper, Professor A M Cormack, Mr R B Coulson, Dr A T G Ferguson, Professor A P French, Mrs Ulla Frisch, Dr L Goddard, Professor Janet Guernsey, Dr Ruth Harper, Sir John Hill, Dr Vere Hole, Dr H G Holland, Professor R W Home, Professor Daphne Jackson, Mr E W Jenkins, Dr G R Lindsey, Dr June Lindsey, Dr Rachel Makinson, Miss Mary Maltby, Dr J B McCaughan, Mr H C Minnett, Miss Ann Phillips, Dr B Rose, Dr C J Sofield, Mrs Hester Sperduto, Dr G H Stafford, Dr F R N Stephens, Miss Mary Stephens, Professor V F Weisskopf, and Dr I F Wright.

I would like to express my special gratitude to Miss Sally Atkinson, Honorary Archivist at the CSIRO Radiophysics Laboratory in Sydney, who has been an enormous boon to me in researching and supplying a variety of information, reports and photographs, relevant to my Australian chapters; and to my husband John, who has been constantly supportive.

Finally I thank Mr Sean Pidgeon who, as Commissioning Editor for Adam Hilger, has been generous in his sound advice and helpfulness to me.

Joan Freeman
Abingdon 1990

Chapter 1

Beginnings

I was lying on my tummy on the upper berth of a small cabin, peering over the edge of the bunk so that I could study the carpet. Light from a circular port-hole illuminated the colours brightly and showed up a complicated and intriguing pattern. There were long zig-zag lines, oddly shaped blobs with curved, tapering tails, and dark jagged strips, interwoven with flowers and leaves on twisted stems. Suddenly, something strange started to happen: the carpet was magnifying itself, and I could see more and more detail as the patch I was studying got bigger and bigger. Evidently the ship must have given an extra lurch, catapulting me from the bunk; but I didn't seem to realise this immediately, being fascinated by the Alice-in-Wonderland behaviour of the apparently self-magnifying carpet. I have no memory of my actual impact with the floor, though I suppose it must have been this climactic event which caused my exercise in pattern analysis to be imprinted indelibly on my mind. This is my earliest recollection. I was four years old at the time. I know this because of the record that in 1922 my parents, with their only child, moved from Perth to Sydney, making the long journey by boat.

It has always been to my considerable regret that the first four years of my life are a complete blank; not a single moment of my Western Australian existence can I recall. And yet the West has a particular fascination for me. I feel somehow rooted in its pioneering history. My father was born in England, and I know little of my paternal forebears; but on my mother's side of the family, whence most of my early influences came, my Australian ancestry spans three generations. My great-grandfather, John Robert North, emigrated from Yorkshire to Adelaide, South Australia, in 1851, when

1

he was twenty-one. Adelaide, never a penal settlement, had first been colonized about 14 years earlier, mainly by liberal-minded English people seeking escape from the social and religious strictures of the period. John Robert is said to have undertaken the enterprise rather than submit to the family tradition which required the second son to become a clergyman. It always excites me to visualize this independent young man facing up to the long and hazardous journey by sailing ship to the other end of the world. He became a farmer, and married an English girl who had emigrated to Adelaide with her parents at about the same time.

A generation later their son, also John Robert, motivated by the same spirit of independence and adventure, migrated to Perth in Western Australia, some 2000 miles away by sea. At that time Perth was a very isolated colony, maintaining stronger links with England than with the rest of Australia. From my grandfather's subsequent accounts I can imagine this vigorous little pioneering town as he first saw it, picturesquely set on a bank of the broad, gently winding Swan River, not far from the port of Fremantle, and the long wild stretches of golden sands which flanked the coastline and submitted to the great rollers of the Indian Ocean. Behind the isolated settlement the hot, dry bush stretched limitlessly, a semi-desert, blazing into colour when its unique wild flowers were in bloom. Without aid from the bigger Eastern colonies life was tough, but Perth possessed a sense of freedom and optimism all its own. Grandpa had a saying, which he was fond of quoting to me in his later years, to illustrate the character of Perth as he remembered it: 'In Sydney they ask you how much money you have; in Melbourne they ask you what family you belong to; in Adelaide they ask you what Church you go to; in Perth they just say "Come and have a drink".'

Here my grandfather started up a successful enterprise, importing textiles from England, and established a "Superior Emporium" in Hay Street, becoming a prosperous and well-respected member of Perth's business community. He married Gertrude Brooke, a young English-born widow, and in due course they had three daughters, the eldest of whom was to become my mother. Her Christian name was Ada, which she detested, much preferring to answer to the nickname of Tommy, apparently derived from her tomboy characteristics as a child.

In 1892, with the discovery of gold in Coolgardie, about 300 miles west of Perth, the great Western Australian gold-rush began, and the fortunes of Perth and the West were transformed. My grandfather, John North, always having a zest for adventure, decided to transfer his interests to gold-mining and developed a very profitable mining enterprise near Kalgoorlie. There he built a family home to which Ada and her sisters came during school holidays. Life on the goldfields must have been tough, particularly for my grandmother who was a gentle and cautious person, less fitted for a rugged way of life than her more adventurous husband. Grandpa, whom I got to know well during his later years, in Sydney, used to regale me as a child with tales of his early struggles in the West. His bright blue eyes, always ready to crinkle into laughter, and his strong, perpetually ruddy features, topped by a generous thatch of white wavy hair, would light up as he described the harshness of the climate; the remoteness of the goldfields; the problems of water shortages; the aborigines who helped on the mines; and the camels, with their supercilious air and curious rolling gait, imported to work under conditions that horses could not survive. I was enthralled by Grandpa's accounts of the exploratory trips he made into the far outback. Once he

Figure 1.1 Ada North (c 1910).

was privileged to go on an excursion with the famous Western Australian explorer John Forrest, a cousin by marriage, and he recalled how, as the food rations ran low, they literally had to tighten their belts against the pangs of hunger. My admiration of Grandpa knew no bounds, although Grandma used to say that he was quite irresponsible where his family was concerned. I have always liked to think that some of his adventurous spirit—but not the irresponsibility—rubbed off on me, lending me an added impetus to pursue, against some odds, my excursions of a rather different type into the unknown world of physics.

Grandpa influence

My mother was sent at an early age to a boarding school in Perth, once being confined there even over a long school holiday period, while her parents made a trip to England. This distinctly lonely environment stimulated her to develop her attributes of resourcefulness and self-reliance, qualities which were to stand her in good stead in later life, and, indirectly, to have a bearing on my early progress. Her school was apparently a good one, and laid particular emphasis on music, allowing Ada to exploit her talents in this direction. She became an outstanding pianist, with an individual and sensitive style, gaining the LRAM Performer's and Teacher's Diplomas, from examiners sent out from England.

Ada had high hopes of a university education through the Adelaide University Extension which had been established in Perth—Western Australia not then having a University of its own. But her ambitions were thwarted by the reversal of her father's fortunes just at that critical time. In the 1900s the great gold boom in the West collapsed and my grandfather's mining venture became worthless. By the time he had paid off his creditors he found himself virtually penniless. Ada took a position in her old school, which allowed her to train as a teacher while earning a small salary, and this she supplemented by giving private music lessons. Her hopes of a university education were dashed.

Then, on the rebound from a love affair which went wrong, she married Albert Freeman, an accountant at a bank in Perth. According to my mother this turned out to be an unfortunate and unhappy alliance from the start. They soon found that they were temperamentally quite unsuited to each other, and had very little in common. My mother was not interested in the banking and commercial world, and my father had no interest in music. The birth of a daughter may have helped a little: I think I was one of the ties that prevented a complete separation at an early stage, though my father must have felt frustrated as he saw me developing into a "mother's girl", to whom my mother was utterly devoted. Regrettably I never really got to know or understand my father. He was fond of me in his way, and his frequent flares of temper were never aimed at me. But I became nervous of him through seeing them directed fiercely at my mother. Perhaps that is why I seem to have a naturally timid nature—a characteristic which has always been a nuisance to me: something to be consciously overcome when necessity arises. And I still cannot bear to see people lose their temper: it is one of the situations that I am unable to handle effectively in dealing with human relations. I do not myself often lose my temper. I remember an occasion when this did start to happen, while I was working with a colleague at Harwell. I normally got on very well with him, but he had just done something so unreasonable that I became utterly incensed. His shamefaced reaction, when he saw me getting angry, was to rush out of the lab and call to some members of our group who were in an adjoining room. 'Hey! Come quickly and have a look,' he shouted. 'Joan Freeman is losing her temper.' They all poured into the lab to witness this unusual phenomenon. The situation was immediately defused; I could only burst out laughing.

When I was approaching the age of four, my father received sudden notice of his transfer to a remote country branch of his bank. According to my mother this happened because he had quarrelled with the manager. He was none too pleased, and my mother was furious, declaring that we could not possibly go because, as she explained to me later, I would not receive a proper education in a country district like that, and she wasn't going to send me to boarding school to be subjected to the kind of existence that she had had to face as a child. So finally my father resigned from the bank, and it was decided that we should move to Sydney, where economic as well as educational prospects looked brighter, to start a new life. Thus it was that, when I was four, we made the sea voyage during which the event that constitutes my earliest recollection occurred—my attempt at analysis of a pattern on a carpet.

I have few notable memories of my first three years in Sydney. There was the episode of the "sleeping doll", a special gift from my mother. It was a large, beautifully dressed example of its kind, with real golden hair, big blue eyes, and eyelids, with long curved lashes, which closed when the doll was laid horizontally. But it appealed to me not so much as an object on which to lavish my affection—I still preferred my battered old teddy-bear—but rather as embodying a new kind of mechanical phenomenon, which invited investigation. How did this eye-closing operation work? I found that I could keep the eyes shut, when I raised the doll from the lying-down to the standing-up position, by holding one of the eyelids down with a finger. Both eyelids did the same thing: they must be coupled in some way inside the hollow head. The experiment of trying to keep the eyes open when the doll was horizontal proved more difficult. How to hold the eyelid up? Perhaps it could be done with a pin. Yes, this seemed to work. But then, suddenly, the force of the pin was too much for the material of which the eyelid was made. A big chip of it broke off. I was overcome with remorse: not primarily for having caused bodily harm to the new doll, though some of my tears were shed on this account; more because I knew that it would upset my mother to see how ungratefully I had disfigured the gift that she had lovingly chosen for me. The last thing on earth I wanted to do was to hurt my mother in any way. However, she must have appreciated this for she was quite forgiving. Perhaps she made a mental note that her daughter was showing signs of a singular interest in mechanics.

An exciting and much-valued possession, which I acquired a year or two later, was a motor car: a beautiful, shiny, bright red affair—a remarkably faithful model of a real motor car—propelled by means of cycle pedals. It had little doors which opened and shut, running boards, mudguards, a horn with a bulb which I could squeeze importantly, and wheels with many thin, silvery spokes. It even had a spare wheel, mounted at the back, on which passers-by used to make surprised and admiring comments, to my great satisfaction. It was indeed a much more elegant creation than the rudimentary objects which pass as toy motor cars for children today. I delighted in driving it around selected footpaths, and coasting dangerously down the hills of the park known as the Domain, which was not far away.

Periodically my mother used to take me to stay for a few days with Grandma and Grandpa, who by then were living in Sydney. They had a florist shop in the suburb of Drummoyne, and lived very simply in the quarters behind the shop. I loved to watch Grandma making bouquets and wreaths, starting with a frame of dried bracken and moss, tightly bound with

Figure 1.2 Upper left: Grandma Gertrude North; right: Grandpa John Robert North; lower left: Father Albert Freeman; right: Mother Ada Freeman with Billy. Photographs taken by the author when about 10 years old.

wire, and building up an artistic assembly by inserting little groups of flowers individually wired to toothpicks. She seemed to be very successful with these creations, which were much in demand. Grandpa looked after the deliveries, and also did most of the cooking. He used to take me on long walks, during which he delighted me with stories of his times in Western Australia, especially during the gold-mining era. He was full of fun, too, and loved to reduce me to helpless giggles, growing red in the face himself with companionable laughter. 'Hello, my sweet potato!' he would say to Grandma, as we returned, breathless, from one of our excursions. She would look somewhat disapproving of his frivolity, whereupon we would both collapse into giggles again.

But my most sustained and treasured memories of my early Sydney days are associated with music and dancing. My mother took me to a School of what was called Eurhythmics and Ballet, run by an entrancing lady, whose name was Gladys Talma. I can see my childhood image of her now, with her long silken hair of palest gold, a delicate, creamy complexion, and an expressive, gentle smile. Her lithe and graceful figure was clad in a white diaphanous garment which added to the ethereal quality of her every movement. I found sheer delight in learning to repeat the sequences which she performed for the small class, and in acquiring the technique of what she called the "follow through"—the linking of successive actions into a continuously flowing expression of the mood and rhythm of the dance.

At home I loved to go through my repertoire, and to invent variations as my mother played her beloved Bechstein piano. She used to play a great deal—an escape, I now surmise, from her marital unhappiness, which I was beginning to sense. I got to know the tone and feeling she imparted to every note of my favourite pieces. I can still hear them in my mind, and can detect how they differ from the interpretations of modern performers. Though prejudiced, I feel sure my mother was an outstandingly gifted pianist. I never tired of composing my own dance routines to match the character of her music: spirited, lyrical expression for Schubert; charm and grace for Mendelssohn; a fairy-tale quality for Greig; and, above all, dramatic, emotional effects for Chopin, whose music I enjoyed most, especially the Impromptus, Polonaises, and the Scherzos. One piece in particular (Scherzo Opus 31) I remember vividly: it began with an inviting upward run of four notes, repeated once; then there were two heavy chords, one in the

base and one in the treble, to which my mother gave tremendous emphasis, followed by a defiant flourish. It made me think
of the building up and crashing down of big breakers on an
ocean beach. I recall the powerful movements which I invented for that sequence.

Those were blissful times for me. I must have had a natural
talent for dancing, since, after a public performance of a ballet
which Miss Talma designed to include representatives from
her various classes, the Sydney Theatre Company, J C Wil-
liamsons, offered my mother a contract for me to appear on
the professional stage. Her refusal was adamant—she had
quite different ambitions for me—though the idea of making
dancing my career appealed to me at the time.

Our shared pleasure in our own very personal world of
music and dance created a strong bond between my mother
and me: an enduring sense of companionship, which
transcended the normal mother-daughter relationship. My
mother never tried to possess me, but tended to live in and for
me, dedicating herself to the objective of doing all in her

Figure 1.3 The author, aged about 6. 'At the time I
rather fancied the idea of dancing as a career.'

mother's motivation

power to keep open for me the opportunities of which she had been deprived, so that I could follow the career of my choice. There were tribulations ahead for her, as my story will relate, but her strong will and singleness of purpose were most important factors in my ultimate achievement of my goal.

That I had been leading a very sheltered life was dramatically demonstrated on my first day at a little kindergarten school in which I was enrolled at the age of six. I felt overpowered by what seemed to be a great crowd of children, all my own age—though they probably only amounted to a dozen or so.

'When is your birthday?' one small girl asked me.
'The seventh of January,' I replied primly. 'When's yours?'
'Mine's the seventh of January too,' she responded.
I was shocked. This was my day—it was special to me. No one else, I felt sure, could claim the same day.
'It can't be,' I said. 'That is my day.'
'Well, it's mine too,' she insisted.
'You're telling lies,' I said indignantly.
'No, I'm not,' she shouted.

My anger was aroused. What a blatant fibber she was! With a deliberate lunge I punched her in the stomach. Her anguished wails summoned the teacher, who, quite unmoved by my declarations of self-righteousness, made me sit outside the class, in disgrace. She must have wondered what sort of little fiend she had acquired, but I, far from contrite, felt very hardly done by. Though by nature a timid child, I was prepared to stand up for my principles, however misguided. It took my mother's powers of persuasion that evening to convince me that there were not enough days in the year to go round for me to have a birthday exclusively to myself.

math!

Settling down somewhat reluctantly to the school routine, I found that I liked doing "sums", for which I frequently got a little gold star stuck in my exercise book as a reward for having got them all right. I was not so good at spelling. However, one day we were being taught the names of the countries of Western Europe; the word "Mediterranean" was mentioned, and I tried writing it down. The teacher was astonished to observe that I had spelt it correctly, and made much of how clever I was. But I only felt uncomfortable and embarrassed, realising that it was just by chance that I had got it right, and that no praise should have been attached. The principle of probability had made its first impact on my consciousness.

Being rather a loner at that stage in my life, I never really took to my kindergarten school. However, the fact that I was not ready to integrate with my school-fellows was probably an advantage to me, as far as my path towards a scientific career was concerned, since I was accordingly not so subject to the "peer pressure" which undoubtedly begins to take effect at an early age: "Little girls do this; little boys do that".

When I was seven we moved to New Zealand for a year, and lived in Auckland. The school I was sent to there was even more of a shock to me than the Sydney kindergarten had been. It seemed huge: an overpowering throng of girls of all ages, accommodated in a number of large forbidding-looking buildings, set in a park-like expanse of lawns, trees and playing fields. The day started with a congregation of the whole school in an "Assembly", during which a series of sombre school prayers, largely meaningless to me, were said. One of them terminated with what sounded like "Deliver us from eagle". I had seen a picture of an eagle in a book, but never a real one, and had certainly not got the idea that the bird was a particular threat to humanity. Was there a plague of them in New Zealand? Or was the school unable to achieve a reasonable perspective on ordinary life? My mother's explanation wasn't of much comfort to me, since the word "evil" was not in my vocabulary and its meaning seemed difficult to grasp.

My next problem was my class teacher, a severe-looking person in a dark tweed jacket and skirt, who was addressed by the girls as "Mister".

'Our teacher is a man dressed up as a woman,' I reported to my mother when I got home. 'It must be so because everyone calls him "Mister".'

'Mister what?' my mother asked.

'Mister Renzy,' I replied.

It transpired that the teacher came from a Channel Island family with the surname "de Renzy", so that what I had heard the children saying was "Miss de Renzy". I somehow never quite forgave her for possessing such an inconvenient name.

Of the learning that Miss de Renzy imparted to me I have no recollection. The only classroom incident I can recall is being knocked out. After standing up at my desk when the teacher entered the room, my seat—a wooden flap which was hinged to the desk behind—tipped up vertically; I was supposed to push it down again before proceeding to sit. Omitting to do this, I shot straight through to the floor, taking, as I descended, a resounding clip under the chin from the back edge of the desk in front of me. I came to in a completely strange place, where I had been deposited and abandoned; the incident left me with a psychological as well as a still evident physical scar.

The school seemed to me very pernickety about small details. For instance, it was considered a disgrace to come to school, as I did one day, without the uniform girdle tied around my middle. To hide my iniquity I kept my blazer on even during the physical exercises conducted on the playing field, in full sun, on a hot summer's day. 'I'm not too hot,' I told the teacher. But ultimately I reached a point of near collapse and had to suffer the ignominy of being stripped of my blazer and laughed at for the evidence thus revealed.

However, as a compensation for the traumas of school life, New Zealand did offer me a number of delights. There was, for example, the discovery of the local Children's Library, where I spent many happy hours on Saturdays and during school holidays. It was here that my lasting passion for reading first developed.

One memory of the New Zealand period stands out particularly vividly in my mind. It was Christmas Day. Having worked my way through my own presents, I set out to investigate what other children in the neighbourhood had acquired. I did not get very far. A few doors down the road lived a small boy of about my own age, whose name was Cliff. I found him sitting on the floor examining the contents of a large shallow box which was labelled "Meccano". It had many compartments containing metal strips in various lengths, shapes and colours, with little holes punched in them at regular intervals. There were boxes with wheels and pulleys, angled pieces, hooks and handles, connecting rods, nuts and bolts, spanners and screwdrivers. And with all this went a book showing how these bits and pieces could be assembled in different ways to make towers, cars, trucks, cranes—there seemed to be no end to the possibilities. My eyes popped in fascination. I squatted down beside the owner of this fabulous apparatus, peering over his shoulder at the book and watching his tentative efforts to join various parts together. I tried to pick up some of the bits, but he promptly took them away

from me. How was I to persuade him to let me have a go? I
decided to wait patiently for a lull in his interest and then to
offer him a positive distraction. I thought of my beautiful
scarlet motor car, which had come with us from Sydney and
was still my pride and joy. Should I suggest that he have a
ride in it? Overcoming a residual reluctance, I decided that
the risk was worthwhile. 'Would you like to try my motor
car?' I asked, at an opportune moment. 'Oh yes,' he replied
enthusiastically. We extracted the vehicle from my house,
and, after strong admonitions to him to take good care of it, I
launched him down the road. Now I was free to double back to
his house and to get down to the wonders of Meccano.

I think that over the next few weeks I gained more pleasure
from that set than did its owner; apparently he was not
particularly mechanically minded. But I coveted very much a
set of my own, and nagged at my mother until she finally
provided me with one. She was no doubt surprised at my
enthusiasm for what was regarded as a boy's toy. My Meccano
set, with its periodic additions, was one of my most valued
possessions for some years, and was frequently used to create
all manner of models, particularly working ones. My mother
provided the upgrading sets up to, but to my regret not
including, the one which added a little electric motor: she was
afraid of my possibly electrocuting myself. This engendered in
me an exaggerated fear of electrical equipment, which I later
had some problems in overcoming. However, my early ac-
quaintance with Meccano no doubt gave me valuable training
in mechanical design and assembly—basic requisites in a
physics laboratory. Years later, especially during the World
War II period, I was to meet Meccano parts again, as regular
laboratory stores items, extremely useful for the rapid construction of prototype experimental apparatus.

It seems sad to me that Meccano has now been superseded
as a children's toy by inferior, if easier to use, plastic devices,
which are not comparable in their versatility or scope for
ingenuity. At the same time, alas, the value of Meccano parts
in the physics research laboratory has also receded, as experimental projects come to be devised on an ever vaster and more
sophisticated scale. Largely gone seems to be the simple
trial-and-error type of improvised experiment. Equipment
nowadays tends to be planned by committee, designed by
professional engineers, manufactured at great expense in time
and money, and finally launched as a virtually unalterable
system. I am glad that I was born into the Meccano era, and

grateful that I was introduced to it at an early age by a little boy who lived in New Zealand.

Chapter 2

Explorations

Soon after our return to Sydney, my parents bought a house in Vaucluse, a suburb about eight miles from the City Centre, straddling a section of the steep and rugged spine which forms the southern arm of the entrance to Sydney Harbour. The west-facing half of Vaucluse, descending precipitously to picturesque little bays edging the harbour, was the "fashionable" side. Its large, distinctive houses commanded superb views across the harbour and up as far as the promontories where the Harbour Bridge was being built. The east-facing, less developed, side of Vaucluse looked out towards the Pacific Ocean, its modest houses clustered mainly along Old South Head Road.

Our house was one of these: a simple but pleasant three-bedroomed bungalow, at the back of which a long verandah gave an uninterrupted view straight out to sea. Behind the back yard there was a sheer twenty-foot drop to the base of a disused granite quarry, which harboured an interesting pool with a thriving population of frogs. Beyond this, the original, rough, scrub-covered terrain, with steep granite outcrops and twisted gullies, tumbled down for a few hundred yards to end in formidable cliffs, against which the Pacific pounded ceaselessly. To the north, the cliffs extended for a couple of miles to the massive South Head with its sturdy lighthouse. For a similar distance to the south of us they skirted a plateau, known as Dover Heights, until they were breached by the bay and famous golden sands of Bondi Beach.

I loved the location of our house, giving ready access to so much wild, unspoiled territory, which constantly invited exploration and make-believe adventures. It was a rich source of wild flowers and a variety of small creatures: grasshoppers; the fat, noisy cicadas; casemoth larvae, crawling about in their large cocoons camouflaged with bits of twigs; and, occasionally, a green mantis with its forelegs raised as if in prayer. From the quarry pool there were tadpoles to be caught and reared in a shallow dish until their legs were sufficiently developed for them to be able to hop out and disappear. They were not popular with my father, however, who was incensed by the croaking of the frogs which he said kept him awake at night.

There were many kinds of spiders too. I learned to be wary of the fortunately rare red-backs and funnel-webs, but there were plenty of other types to observe: I enjoyed surprising a trap-door spider and watching him scuttle back to his silk-lined burrow in the soil, pulling his cunningly constructed lid down behind him so that all trace of his lair had disappeared. There was also an interesting kind of web-building spider who liked to make a shelter for himself by curling up a dead leaf and fastening it to the centre of the web, where he could lurk undetected. I once found an enterprising member of this species using a paper bus ticket in lieu of a leaf. The small rectangle of thin red paper must have blown into his territory from the vicinity of the bus terminus, which was not far from our house. I marvelled at the ingenuity of this little creature, typical of the resourcefulness of nature.

Our view of the sea was also full of interest, one of the memorable sights being the red orb of the sun popping up out of the sea at dawn. From our verandah I could observe the ships—big passenger liners, and cargo vessels of all types—plying back and forth along the coast to and from Melbourne, or occasionally appearing over the horizon, presumably from New Zealand. It was usually just opposite our house that the pilot boat came out to meet the bigger ships. In calm weather it could pull adroitly alongside, just long enough to allow the pilot to get on or off. In rougher seas he had to be taken across in a small tender. Sometimes the sea was so wild that the pilot boat just had to lead the vessel in. Easterly winds were occasionally strong enough to lash sea spray right across the few hundred yards of bush to our house, covering the windows with salt.

Swimming with one of my friends was another delight. On Saturday mornings we would go down to Rose Bay, on the harbour side, where a large rectangular section of the water was fenced off as a swimming bath, safe from sharks but fed continuously by the waves and tides. There I learned to swim and dive like the proverbial fish.

Those childhood days in Vaucluse were particularly happy and carefree, stretching out into what seemed like a perpetual summer. I enjoyed my school too. I went to the Sydney Church of England Girls' Grammar School (popularly known as SCEGGS), which was located near King's Cross, a busy five-way junction about a mile short of the City Centre, and, in those days, a quite salubrious area. In contrast to my previous experiences, I took to this school immediately, and grew more and more attached to it with the years. The main building was a comfortable-looking old stone mansion, to which had been added various annexes and a large gymnasium. The grounds were not extensive but were broken up into separate, attractively laid out areas by mature trees and shrubs. The classes were small and intimate, and the teachers kindly, giving me the feeling that I was an individual, worthy of personal consideration.

The spirit of the school appealed to me too. Christian morals and principles were strongly upheld, without coercive religious attitudes. The inspiration derived partly from the well-founded school traditions, but mainly from the vigorous and dedicated personality of the Headmistress, Miss Dorothy Wilkinson. She had long fair hair, drawn back in a big bun from her characteristically upturned face, and penetrating blue eyes which commanded respect and obedience. But there was at the same time a gentle and understanding side to her nature. I looked up to her at first with awe and trepidation, but, as I got to know her, with admiration and warm affection. I was also to have particular reason for personal gratitude to her.

Being young for my class, and small for my age—I barely exceeded five feet when fully grown—as well as being shy, I was initially subjected to a certain amount of teasing from some members of the class, and was often referred to as "The Squib". However, as my intellectual abilities, and my willingness to help others with their maths homework, came to be *Eng + math* recognised, the teasing turned to respect, and I began to make friends and to gain self-confidence. Increasingly I enjoyed the lessons, especially English and maths. Even English grammar—parsing and analysis—which seemed to be a bugbear of most of the class, appealed to me with its logical procedures; and the freedom to compose "essays", stories, and poetry, was a delight. One year we had a particularly stimulating young

Figure 2.1 Dorothy Wilkinson, Headmistress of the Sydney Church of England Girls' Grammar School, Darlinghurst. (Courtesy of SCEGGS)

English teacher who encouraged me to stretch my imagination and creative capacities. But then came the morning when she staggered in with what was obviously a gigantic hangover. Alas, that was the last we saw of her. Maths, however, was my favourite subject; I still recall the thrill of my introduction to algebra.

The nearest I got to a science subject during my first few years at SCEGGS was botany. But I found the going rather slow and monotonous—except for the occasional experiment, like growing bean seeds, wedged with damp blotting paper against the glass wall of a jam jar: whichever way round the beans were placed, the roots could be seen always to find their way downwards and the shoots, upwards. It was not until some years later that we started to learn chemistry, which I found more interesting and challenging.

But it was at home that I first began to read and absorb

general scientific topics. This came about through my acquisition of what proved to be my most valued, and indeed most important, childhood possession. I was about ten years old when my parents bought for me the ten volumes of Arthur Mee's *The Children's Encyclopedia.* I recall vividly their arrival one day in a huge box, and the excitement of unloading each volume, with its handsome dark blue binding, richly embossed in gold. My father, who was adept in simple carpentry, made a special bookcase for them, and this treasured unit was my window onto the delights and mysteries of a new, wide, world. Reluctantly, I had to leave it behind when I left for England in 1946, but, browsing recently through the set which I located in the Cambridge University Library, I found that the copious photographs and illustrations as well as much of the text in the 7000 odd pages were still gratifyingly familiar to me; it was like suddenly meeting again a cherished childhood friend.

What a debt I owe to Arthur Mee for his very substantial contribution to my early education, particularly in the scientific field, to which I otherwise had little access. Published in 1926, the Encyclopedia was quite up-to-date when I first met it, and though it now has an old-fashioned appearance, and some of the scientific facts presented in it have been superseded, the essential spirit is as relevant as ever, and I believe it surpasses most of the more modern equivalents.

I absorbed eagerly the chapters on literature, the arts, history and geography, but found of overriding fascination the scientifically oriented chapters: vividly and imaginatively illustrated sections on natural history, evolution, astronomy and the solar system, the formation of the Earth, and the nature of the matter of which it is composed. Here, I think, began my conscious interest in science. There was also a skilfully presented set of chapters under the heading of *Wonder*, in the sense of a desire to know, in which a thousand and one questions which a child might ask (like: What makes a rainbow? Why does smoke go up the chimney? Does the air ever get used up? What is a mirage?) were answered in terms simple enough for me to follow. In this way I was introduced naturally to many qualitative features of the physical sciences, such as gravity, properties of light, the laws of motion, chemical compounds, waves, and so on. This was to stand me in good stead as a basis to build on when I made my rather late formal entry into the field of physics.

In another group of chapters—sensitively presented essays

on ideas and emotions of the mind—I found defined the
concept of *Wonder* in the sense of marvelling at: the conscious
mental reaction of mingled surprise, admiration and excitement which can be evoked by the perception of a natural
phenomenon or a special product of human creativity. The
whole spirit of Arthur Mee's Encyclopedia seemed to promote
for me this sense of wonder.

The Encyclopedia also provided me with an almost unending source of entertainment out of the chapters on "Things
to make and do". Passing over the sections on embroidery and
needlework, most of which I thought rather dull, I rejoiced in
the puzzles, games, and models to make, and, above all, the
"scientific" experiments. I learned how to make a syphon; how
to detect the rotation of the earth; how to observe atmospheric
pressure; and how to make water rise up inside an inverted
glass by first heating the air inside it. I discovered how little
pieces of camphor could be made to drive a paper boat about
on the surface of a dish of clean water, and how a match could
be ignited with the aid of a magnifying glass in the sun; in
fact I found that the Australian sun, unlike Arthur Mee's
weaker English sun, was strong enough to allow me to burn
pictures and writing into a sheet of paper. One day I made the
exciting discovery that my mother's dressmaking scissors were
magnetic, and could be used to turn pins and needles into
little magnets by stroking them across one of the scissor
blades. I astonished the local chemist once by asking him for
red and blue litmus paper so that I could test the acidic and
alkaline properties of vinegar, washing soda, and other
domestic substances. But my experiments were not always
successful, and my mother became agitated at times when I
wreaked havoc in her kitchen.

In most of the Encyclopedia's illustrations for the scientific
experiments, a boy rather than a girl was drawn. I noticed
this, but thought little of it. After all, I had already discovered
that story books for boys, like *The Boy's Own Annual*, were
more exciting to me than those written for girls, so, although
most of my friends were girls, and I enjoyed their companionship, it didn't seem surprising to me that I should prefer boys'
scientific experiments to girls' embroidery.

However, there were some problems for me in experiments
which required a collaborator. I remember once trying to
persuade one of my girl-friends to help me in a study of blood
flow. The idea was for me to lie centrally along the horizontal
plank of a see-saw, so that it was exactly balanced. She had to

hold up in front of me a soothing picture which would make me feel completely relaxed. Then she was to substitute for the picture an arithmetical problem, previously set out on a piece of cardboard. The effort of this mental exercise was supposed to send an extra supply of blood to my brain, so that, according to the convincing illustration in the Encyclopedia, the see-saw would become unbalanced and my head would tip downwards. However, what with the agitation involved in getting into the balanced position and making my friend understand what she had to do, to say nothing of the distractions caused by curious onlookers in what inevitably had to be a children's playground—for where else could one find a see-saw?—the flow of blood to my brain had no doubt reached saturation point well before the experiment began.

'I can't see any difference,' grumbled my friend, after I had struggled for a minute or two with a task like multiplying 26 by 43. 'Perhaps if you didn't breathe so hard it might work better.'

'Well, how about you trying it?' I asked. But by then she had stamped off, muttering that it was a stupid game anyway.

After that fiasco I tended to concentrate on experiments which did not require an accomplice. However, the telephone project, which did need cooperation, proved to be great fun. Telephones were a novelty at that time, and fairly rare in private homes. The idea I had read about was to take two empty jam tins, punch a hole with a nail in the base of each, and thread the ends of a long piece of string inwards through the holes; the ends were then knotted so that the string could not be pulled out of the tins. The two boys who lived next door but one to me were participating in this exercise, which was being conducted in my back yard. One of the boys held a tin over his ear while I, pulling to make the string taut, spoke into the other tin. We could then reverse our roles so that he spoke and I listened. Amidst some confusion and argument we tried to compare the sounds we heard with and without a tin held over the ear; but it was difficult to tell whether the "telephone" was really working.

'We need to be further apart, so that we can't hear each other except with the telephone,' I decided. 'Why don't we make the string long enough to reach from my verandah to yours and try it that way?'

The boys were enthusiastically cooperative. We pooled the resources of our respective mothers, as far as collecting enough string was concerned. I hoped that all the knots would not spoil the effect. Then we had to negotiate with the next-door neighbour, who was digging in his back garden, for permission to take the string across his yard. A large man, with an untidy shock of light brown hair and a receding chin that reminded me of a shaggy dog, he was fortunately quite easy-going. He had a habit of quoting Shakespeare in a deep, melancholy voice while he was digging, and we could hear him declaiming: 'Friends, Romans, Countrymen, lend me your ears', between thumps as he drove the spade into the soil.

'Mr Boland,' I called, over the fence, 'do you mind if I come in just to lead a piece of string across from my verandah to Bill's and Jack's on the other side of your yard? It's just for a little while, to do an experiment.'

'Go ahead,' he said, amiably enough.... 'I come to bury Caesar, not to praise him.'... Thump!

We found that, by leaning over the railings of our respective verandahs, we could keep the string taut while speaking into or listening to the tins. There was much shouting while we tried to coordinate our operations. An exchange something like the following ensued:

'Hullo.'
'Hullo.'
'Are you talking into the tin?'
'No, not yet.'
'Try it now.'
'Hullo.' The sound could have been heard a quarter of a mile away.
'Not so loud.'
Silence, except for: 'The evil that men do lives after them.' ... Thump! ... 'The good is oft interred with their bones.'
'What's happening?'
'Jack says the string is touching a verandah post.'
'Alright, lean out a bit further.'
'Hullo. Can you hear me now?'
'Yes. Is that into the tin?'
'No, but this is.'
'Try talking more quietly. And perhaps we should say something different into the tin from outside it, so we can compare. Say "Hullo" outside the tin, and "Oranges" inside.'
Silence.... Then: 'Come I to speak in Caesar's funeral.' ... Thump!
'What's happening now?'
'Jack's gone to get his bell.'

'Why?'

'He says all telephones have bells.'

'Oh, let's get on with it. Let me try while you listen. Hullo ... Oranges ... Hullo ... Oranges ...' with decreasing intensity. 'Can you hear a difference?'

'Your voice sounds more muffled in the tin. You listen while I try.'

'And Brutus is an honourable man.' ... Thump!

Finally we decided that we had gained reasonable evidence for the success of our telephone. Well satisfied with our efforts, the three of us repaired to my verandah and celebrated in lemonade.

I have to admit that the boys were a better bet than my girl-friends for this kind of enterprise. However, when it came to gentler and more imaginative pursuits, I found that girls won hands down over the boys. The girls were prepared to help me make little dens, or to play make-believe games in the wild scrub along the cliff tops. They mostly drew the line, though, at the games I invented which involved writing or arithmetic. These I pursued happily enough on my own. Writing stories was a particular passion; one of these, for which I won a prize, was my first publication—on the Children's Page of the Sydney Morning Herald.

Chapter 3

Where There's a Will

In 1929, when I was eleven, my carefree childhood days were abruptly disturbed. The Great Depression hit Australia, very hard. The country's economy was plunged into chaos. Businesses of all kinds were collapsing, and unemployment was becoming rife. I could sense the immediacy of my parents' apprehension. Then, one day, their fears were realised when my father came home announcing that his financial firm was closing, he was losing his job, and would be unlikely to get another one. It was the beginning of a disaster period not only for us but for the whole country. At the height of the depression one working man in three became unemployed. Professional men such as doctors and lawyers were said to be trying to get work as road-diggers.

Grimly my parents took stock of our situation: most of their capital was invested in the house, which was heavily mortgaged. There was something called a second mortgage through which they were required to pay only the interest on the borrowed money, but this alone was a serious drain on their resources. My mother, who had been taking private music pupils for some time, found that many parents could no longer afford to pay for this extra for their children. Her income from this source barely covered my school fees.

My father started hinting that I should go to a State school, which would be free, but my mother would not hear of it. She felt that to concede to such a move would be a betrayal of her most cherished objective: to see that I had the best possible educational opportunities. The idea was ingrained in her that the private schools of Sydney (the equivalent of the Great Public Schools of Britain) offered a better all-round training

24

— for life as well as academically — than the bigger "mass-production" State schools. She was also, true to her upbringing and the social attitudes of the time, convinced that her daughter should grow up within her own social class, as represented at SCEGGS and other private schools. Above all, she was influenced by my strong feelings in the matter. I was by then devoted to my school, and felt very much a part of it. I enjoyed my lessons, generally admired the teachers, and had become strongly attached to the friends I had made there; my self-confidence was growing, and I was overcoming my natural shyness. The thought of a large, impersonal State school, in which I would have to struggle to establish myself, was quite repugnant to me.

So my mother sought all possible means to maintain me at SCEGGS. While my father began growing all our vegetables, she took in a paying guest — a young bachelor whom I found quite entertaining; but, when he started making unwelcome advances to my mother, he had to go.

Then, one day, my mother saw an advertisement from a lady who was moving to another district and wanted to sell the "goodwill" of a small kindergarten school for five and six year olds, which she was running in her rented home. On investigation, the proposition looked unattractive in a number of ways, but the rent was low, and the house was only a five minute walk away from ours. So my mother decided to take it on.

Opposite our house, on the other side of Old South Head Road, a rough, zig-zag path, weaving its way steeply past massive boulders, scaled a natural escarpment of some fifty feet, from the top of which rose a short stretch of metalled road, known as Clarendon Street. Straggling up the left-hand side of this road were about half-a-dozen small bungalows; the one at the top was where the kindergarten lady and her husband were living. Shortly before they moved out, I came with my mother to see the place, little realising what a significant part it was to play in her life, and, indirectly, in mine.

Number 1 Clarendon Street has since been rebuilt as a comfortable home, but it was then a mean little house, with a tiny front garden dominated by a battered palm tree. Steps led up to a small verandah from which the front door opened straight into the living room. This was the only room of any size in the house. Part of it was given over to the kindergarten, and was equipped with a few little tables and chairs. To the right was the kitchen, with one door leading to the outside of the house, and another opening into a diminutive wash-house, barely large enough to contain a typical pair of stone troughs, a copper gas boiler and a wringer. Opposite the front door another door led from the living room into a short passage, giving access to two small bedrooms, and a bathroom just wide enough to accommodate a tub and washbasin. At the far end of the bathroom was a door leading, via a flight of steps, down to an unkempt back yard. Here, about thirty feet from the back door, a little wooden hut housed the W.C. Although by then connected to the main sewage system, it was still in the position it had had to occupy, at a statutory minimum distance from the house, when it had been a primitive wooden seat surmounting a metal can which was regularly changed by the night man. So, whatever the weather, all the occupants of the house, including the kindergarten children, had to trek across a stretch of rough and often muddy grass every time they wanted to visit what my mother termed "the inconvenience".

Besides the obvious drawbacks of this arrangement, it led to the floors of the house frequently getting very messy, as my mother soon discovered. Her first job in the afternoons, after the children had gone home, was to scrub the floorboards with a bucket and mop. She could not afford to pay a cleaner.

The kindergarten had been given the unaccountable name of Neuchalet, which my mother retained because of the so-called "goodwill", although she found that this was largely illusory. There were few pupils to start with, but they increased steadily in numbers as the reputation of the school grew. Soon she was able to introduce a number of improvements, such as having the back yard covered with a layer of macadam as far as the lavatory, and acquiring some new furniture and a cheap piano.

The parents seemed to like my mother's individual style of handling and teaching the children, and, as time went by, a number of them asked her to keep their children on as they got older. A few new, somewhat older girls also came, particularly in cases where the parents felt that their offspring needed individual care or had not been treated satisfactorily at their previous schools. After a while my mother was able to employ a young woman to look after the youngest children while she concentrated on the older ones, who were accommodated in the two small rooms at the back of the house. Her only refuge then was the tiny wash-house off the kitchen. A board across the top of the troughs constituted her "desk", and, if she wanted to sit, one of the little kindergarten chairs could just be squeezed in. I sometimes squeezed in with her, if I was about. Her spirit was remarkable: she would joke about her "office" while enjoying a thankful break, however small, from the continuous demands upon her.

The school was making a profit, though the income hardly met the combined demands of interest on our house mortgage, my school fees, housekeeping costs, and other expenses for the three of us. Once, when the financial situation became particularly tight, one of the mothers absentmindedly paid her term bill twice. Tantalized, my mother phoned her to tell her of her mistake. 'Oh, don't bother to send it back,' she said. 'It will do for next term.' She did not realise what a godsend that little bit of cash in hand was to my mother.

The situation was really frustrating. With a little capital my mother could have moved the school to a bigger, more appropriate house, expanding it on the strength of her growing reputation for good individual teaching and for inspiring her girls to give of their best. But she was stuck: the average potential parent could hardly be attracted by the appearance of a poky little bungalow with its W.C. half-way down the back yard. Thus she found herself with only a few pupils in each grade, and, since they wanted to stay on with her from year to year, she was teaching a variety of subjects simultaneously at different levels, up to a higher and higher grade. She had to work painfully hard to keep abreast of this expanding curriculum.

In spite of all her struggles the state of my mother's finances gradually worsened until it became desperate. Some expenditure had to be cut. It seemed that, after all, I would have to leave my beloved SCEGGS and go to a State school. We were plunged into gloom as, with a heavy heart, Mother went to see Miss Wilkinson, the SCEGGS Headmistress.

That evening, she returned transformed. Miss Wilkinson had said that I was such a good and promising pupil, likely to bring academic credit to the school in the public examinations, that they could not afford to lose me. So she decided on the spot that, by a special arrangement which she, the Head, would personally make without anyone else needing to know about it, all my fees would in future be waived. Dorothy Wilkinson, with characteristic tact, had evidently persuaded my mother that this was no case of charity, but rather a business proposition to benefit the School. So my mother came home in triumph. Both highly gratified by the turn of events, we threw ourselves with renewed vigour into our respective tasks. I now felt an added responsibility to do well—to justify the Head's faith in me.

Our straitened circumstances placed severe restrictions on our way of life at home, although my mother did everything in her power to shield me from the worst effects. Holidays were out of the question for her and for my father; however, they became a reality for me, thanks to the kindness of some of my school friends and their understanding parents. One family in particular, that of the local doctor, not only took me on holiday with them, but also adopted me into their large family circle, becoming an important influence for which I have always been grateful. Dr Stephens, a high-principled, humanitarian man, and Mrs Stephens, his gentle, warm-hearted wife, had six children. Their one boy, Dick, was two years older than I; Mary, one year older; and Ruth, slightly my junior. There were three younger girls, Elizabeth, Joan, and Margaret. Introduced into the family by Mary and Ruth, who were at SCEGGS, I delighted in sharing their enthusiastic, at times quite boisterous, way of life, in their big house at the top of the hill in Vaucluse. I learned the principle of give-and-take necessary in a big family; joined in their pranks; shared their individual joys and problems; participated in their communal activities, like Saturday afternoon tennis on their grass court; and discovered with them the wonders of the Blue Mountains, where Dr Stephens used to rent a house during school holidays. I even acquired a little science from Dick, who had a passion for building radio sets. The Stephens family was a great boon to me, enriching my early life, and helping me to overcome the inhibitions characteristic of an only child.

All the time my scientific interests were developing steadily. At school I was learning chemistry, and, although I suspected that our teacher was not very good, I found the subject exciting and absorbing. Our classes were held in a fine new laboratory, with two long benches carrying a number of individual sink units each with water and gas outlets and a bunsen burner. I longed to be able to use them for experiments, but the teacher always demonstrated the most interesting ones at her own elevated bench. I think she was afraid that we might burn ourselves or blow ourselves up.

Figure 3.1 The Stephens family, early 1930s. Left to right, front row: "Nursie" (Miss Malcolm), Joan, Mrs Stephens, Dr F G N Stephens, Margaret, Ruth. Back row: Mary, Dick, Elizabeth.

For the year 1931, the prize I received for being top of the class was, at my request, a book about science. Its title was "*Science for All*". I found it rather heavy going: its half-dozen articles were written by eminent scientists, mostly Professors with lots of letters after their names, and must have been intended for more mature readers. But I was not deterred. The chapters on physics and astronomy particularly fascinated me, and I felt impelled to try to fathom out the presentations on the basic physical laws and the properties of matter, as well as discussions on recent developments.

"Science for All"

physics + astronomy

In 1932—I was fourteen by then—an event occurred which was to prove of special significance for me, as well as being a dramatic milestone in the advance of the science of nuclear physics. One morning my mother, in a hurried perusal of the newspaper, noticed an item which she thought might interest me. It was a small paragraph—still etched in my memory—near the bottom of the second column from the left on the front page of the Sydney Morning Herald. The headline read:

SPLITTING THE ATOM
AT THE CAVENDISH LABORATORY

1932

There followed a brief statement about an experiment by
Drs Cockcroft and Walton at the Cavendish Laboratory, Cambridge, in which they had succeeded in splitting the atom
using a very high voltage. The account was incomprehensible
to me, but the impact was great: in my chemistry lessons at
school I had been taught the definition of an atom as "the
smallest quantity of any element; *it cannot be divided.*" In
some excitement I took the newspaper with me to school and
showed it to our teacher during the chemistry class. She read
the item blankly, evidently thrown off balance by this sudden
challenge to the authority of her teaching, and was at a loss
for an adequate response. No doubt she was silently cursing
her precocious pupil for embarrassing her in front of the class.

But the event was a revelation to me. Suddenly I realised
that facts laid down in textbooks were not necessarily inviol-
ate: through scientific research, discoveries could be made
which revised or even contradicted existing beliefs. To a
young schoolgirl, used to accepting as absolute truth printed
definitions and the pronouncements of the teachers, this was a
revolutionary new concept. How exciting it must be, I
thought, to be involved in research work of this kind. The
extent of the impact on me of that event was brought home to
me recently when, to satisfy my curiosity about the accuracy
of my memory, I looked up the old newspaper report of some
fifty years ago. In the extensive London Newspaper Library,
copies of the Sydney Morning Herald published in the 1930s
are available on microfilm. In the issue for May 2nd, 1932, I
found, on the front page, near the bottom of the second
column from the left, the report on Cockcroft and Walton's
splitting of the atom, exactly as I had remembered it.

At the end of 1932 I sat my first public examination, which
was called the Intermediate Certificate Exam—roughly equivalent to the English 'O' level. I took eight subjects and had
ambitions to achieve eight A's—a somewhat rare feat at that
time. In the event, however, I gained seven A's and one B,
which seemed to satisfy Miss Wilkinson. The B was for—of all
subjects—chemistry. This disappointed me, though I was not
too downcast—feeling that the standard of teaching had been
at least partly to blame. My enthusiasm for the physical
sciences remained undiminished.

I now faced a serious dilemma. Moving up into the class
immediately before that of the final year—the year of the
Leaving Certificate (Matriculation) Examination—I found

that the school, in keeping with virtually all the private girls'
schools in Sydney, offered no further chemistry, and had
scarcely even heard of physics. The only science subject they
taught was zoology. This, I discovered, was interesting in its
way, but the excitement to be gained from dissecting frogs
and cockroaches, in order to examine their circulatory systems
and alimentary canals, was limited as far as I was concerned,
and learning the names of all the bones in a rabbit's skull I
found positively boring. English, French, Latin, and above all,
maths, I enjoyed, but I yearned to pursue the physical sciences.

My sights were set on a science course at the University,
and in this ambition my mother encouraged me. 'We'll be able
to get you there somehow,' she kept on saying, with a
determination that seemed to transcend all her difficulties.
'You'll do well enough in the Leaving Certificate Exam to get
a bursary, and perhaps a Scholarship.' It was unthinkable to
her that I should fail to have a University career, as she had
done, because of financial circumstances. She did not mind
what subjects I chose, so long as I had the opportunity of
getting a degree.

The immediate problem was that, in order to maintain the
option of taking physics and chemistry as my major subjects
at the University, it appeared to be highly desirable, if not
essential, that I should be studying at least one of them now.
A parent of one of Mother's pupils, who was the head teacher
in French at one of the big high schools for boys, was
knowledgeable and helpful in his advice: the boys who aimed
to follow a course in the physical sciences at Sydney University mostly took both chemistry and physics to Honours
standard for the Leaving Certificate. "Scottie" (his name was
George Scott) considered that, if this was my aim also, it was
necessary for me to gain a pass in at least one of them, along
with maths, in the Leaving Certificate Exam.

Here, then, was my dilemma. We had observed that, in the
Leaving Certificate results for a few of the major girls' high
schools in Sydney, one or two girls passed in chemistry each
year, although passes in physics were rare. So, once again, the
possibility of my changing to a State school was discussed: in
this way I could perhaps take chemistry. But, as before, I was
very unwilling to contemplate such a move. Apart from the
sense of moral obligation to stay at SCEGGS, in view of what
Miss Wilkinson had done for me, I felt that a change at this
stage would be so disruptive that I would probably not do well
in the exams anyway.

night school

I'm not sure who it was (Scottie, perhaps) who first sowed in our minds the idea that I might try to go to the Sydney Technical College for evening classes in chemistry or physics, while carrying on at SCEGGS. The Sydney Tech, we discovered, was intended for people, such as apprentices in engineering firms, who had left school but needed to gain some extra training and qualifications to further their careers. Courses in chemistry and physics were included in the curriculum. Seizing eagerly upon this idea, my mother immediately made an appointment to see the Head of the Chemistry Department at the Sydney Tech. We decided that it would be better, from the exam point of view, for me to try for chemistry, of which I had already learned something at school, rather than physics, in which I had had no grounding.

The Sydney Technical College was situated in a district called Ultimo—aptly named, I thought, as the ultimate in dubious, shabby-looking old buildings and mean terraced houses—a short tram-ride from the Central Railway Station. The main building of the college, of uninspired red brick, Victorian style architecture, was set back slightly from a low wall which skirted the main road. The office of the Head of Chemistry was bare and fusty, furnished with shelves laden with papers and books, and a small untidy desk at which sat an oldish, heavily built, white-haired man. He received us with a show of heartiness, removing the papers from a couple of upright wooden chairs so that we could sit down. He listened impassively to my mother's story of how I came to want to do chemistry at the Tech. He questioned me very little, and I could feel a barrier building up in him. His response was forthright.

'This College is for the technical training of apprentices and the like,' he said. 'It is no place for a young schoolgirl. Your proposition is not feasible.' Standing up and ushering us out peremptorily, he patted me on the back in what he probably thought was a fatherly fashion and urged me to forget about subjects like chemistry. 'Go back to your domestic science,' he said, 'that subject is more suitable for a girl.'

I felt shattered. My mother was furious, but not yet beaten.

'Come on,' she said, 'let's go and see the Head of the Physics Department.'

After some searching we found a door labelled "G H GODFREY, PHYSICS DEPARTMENT". A gentle 'Come in' answered our knock. The room was similar in appearance to the one we had just left, but the man behind the desk was very different. Youngish, fairly small, with curly brown hair, he was not particularly good-looking, but had a responsive face and a quiet, kindly manner. Having recovered from his initial surprise at our sudden appearance, he listened attentively as my mother went through her story again and told him of our abortive interview with the Head of Chemistry. He betrayed a flicker of a smile as she recounted indignantly what the latter had said. Perhaps he doesn't altogether like the Head of Chemistry, I wondered. He then began to question me closely, probing to discover how much I knew, particularly in maths; what I had gleaned about physics from various sources; and why I was so keen to pursue the physical sciences. I sensed that he was warming to me, and felt encouraged. He then explained to me what attending his classes at the Tech would be like: the classes were large, virtually all men; they were tough-looking characters, appreciably older than I was; I would/have to work on my own in the practical classes; I would probably not find the going easy. How did I react to all that? he asked.

With growing excitement I said that I understood and

Figure 3.2 G H Godfrey, late 1950s. (Courtesy of the University of New South Wales Archives)

accepted all his points, and was very keen for him to take me on. He seemed pleased with my response. Turning to my mother, he explained that they had never had a schoolgirl in the physics classes before, but that he admired my spirit and was tempted to give me a chance. 'You must understand, though,' he continued, looking at her intently, 'that I am taking some risk, of which the authorities might not approve. This is not a safe area for a young girl on her own. I will take Joan on provided you promise always to bring her to the classes and to collect her afterwards to take her home. The classes run from five o'clock until seven. If you wish to wait here for the two hour duration of each class, I can find some unused room for you to sit in.'

Oh, my poor mother, I thought. But my heart leapt at the opportunity I was being offered. My mother agreed without hesitation: she was willing to do anything to further my career. Mr Godfrey seemed satisfied with her assurances. After showing her a bare and dismal-looking room which she could occupy during my classes, and pointing out the lecture room to me, he saw us out cordially, saying that he looked forward to seeing me at the classes which were to begin shortly.

My mind was in a turmoil. Uppermost was the thrill of achievement—an acceptance into a Sydney Tech class, snatched out of the jaws of apparent defeat. At the same time, I couldn't help feeling apprehensive of this strange and somewhat scary environment in which I would have to make my way. I also realised the enormous extra drag this enterprise would impose on my mother, who was already overburdened with work and could ill afford to give up so much of her time to me in this way. But I could see that she was buoyed up with triumph. She assured me that it would all work out quite satisfactorily: she would bring a load of the children's homework books with her from school; they had to be corrected anyway. A great wave of gratitude surged through me as I thought what a marvellous mother she was, as well as a delightful companion.

After school, on the big day, I stood at the tram stop at King's Cross, waiting for the particular tram from Vaucluse that my mother had managed to catch, lugging along her heavy suitcase full of school books. There she was at last, waving cheerfully. In the City Centre we changed to another tram which rattled its way downwards—gravitationally and environmentally—to Railway Square, and then on to Ultimo.

Once again we entered the Sydney Tech Physics Department. Leaving my mother in her lonely room, I proceeded in trepidation to the lecture theatre.

The room was large and formidable, with tiers of wooden benches and well-worn desks rising steeply towards the tall dark ceiling. Scores of young men were shuffling in, climbing the staircases on either side, and sliding along the seats. If they felt any surprise at the sight of an undersized girl, standing hesitantly in her school uniform with her hair hanging in two pigtails, they gave no intimation of it. I perched myself at the end of one of the rows, near the back, leaning forward in order to reach the desk fixed in front of me. Now I understood why the lecture theatre was tiered so steeply. Small as I was, I could see right over the top of the heads in front of me, gaining a clear view of the lecturer's bench and the blackboards lining the wall behind it.

Then Mr Godfrey came in and the general hubbub subsided. As he began to speak I quickly became oblivious of my strange surroundings, concentrating intently on this, my first physics lecture. He talked slowly and firmly, starting with an introduction on what physics was about, and then going on to discuss the basic units of length, mass and time, the measure- *Lecture* ment of physical quantities, and the question of accuracy in experiments. I remember well his demonstration of the use of Vernier calipers, which he said we would be meeting in our first practical class. This was an ingenious instrument which could measure the dimensions of small objects to an accuracy of one hundredth of an inch. As the lecture ended, I realised with relief and satisfaction that Mr Godfrey's expositions had all been clear to me, and that his standard was going to be just about right.

After the lecture we were directed to a laboratory for the practical class. It seemed a huge room, well filled with rows of long benches and stools. The Demonstrator, Mr Price, was a *Lab* pleasant-looking man, with a round, open face, but also a firm, no-nonsense air. He had evidently been told about me by Mr Godfrey, since he immediately greeted me with an encouraging smile and asked me if I would mind waiting until he had dealt with the other students. He instructed them to get together in pairs, and went round issuing Vernier calipers and a little brass cylinder to each pair, telling them to make and record a series of measurements of the length and diameter of the cylinder, and to work out the mean values. He then returned to me.

'I'm afraid I've run out of Vernier calipers,' he began. I must have looked crestfallen, for he smiled gently as he continued: 'Don't worry. I've brought you an instrument called a Micrometer Screw Gauge which is much superior to the Vernier calipers; it is twenty-five times more accurate.' He held out a U-shaped instrument with a spindle graduated in millimetres, and pointed out how the object to be measured could be squeezed between the jaws of the "U" by a rotating graduated thimble.

Feeling somewhat bewildered, I examined the micrometer, struggling to understand how it worked. Then suddenly its functions became clear, and I could see that it was capable of an accuracy of one hundredth of a millimetre. My spirits rising, I proceeded to take measurements of my brass sample. By the time Mr Price returned to see how I was getting on, I had a record of my series of measurements entered in my new laboratory notebook. He seemed genuinely pleased, and made encouraging comments, which gave me enormous satisfaction. I have often wondered whether he gave me that micrometer deliberately in order to test my capabilities. He proved to be a good friend and ally throughout the course, so that I never really suffered from the lack of a partner in the practical classes. In fact, the need to be largely self-reliant and resourceful was an excellent training for me.

Well satisfied with the way my first session at the Tech had gone, I went off to pick up my mother, who was still labouring with her school books. She could see that I was happy and elated, and her tired eyes immediately lit up with pleasure. The journey home passed quickly as I gave her an enthusiastic and detailed commentary on everything that had happened.

The Tech classes continued to go well and I became more and more absorbed in, and delighted by, my studies. The recommended text book was a great help. It was a massive, satisfyingly important-looking volume, entitled *Physics: Fundamental Laws and Principles*. The authors were Edgar Booth, Lecturer in Physics in the University of Sydney, and Phyllis Nicol, Tutor in Physics at the Women's College of the University. The second author was a woman, I noted with interest, wondering if I might some day meet her. The book was well set out, though somewhat lacking in descriptive material. It had for me the enormous advantage of including completely worked-out answers to many of the problems which were appended to each chapter. This self-teaching

quality helped to make up for my lack of basic early training
in the subject.

Sometimes I came up against difficulties in comprehension
that were not resolved either by Mr Godfrey's lecture or by
Booth and Nicol. Students used to line up at the end of Mr
Godfrey's talks, to ask him questions, but I was always too
shy to muscle my way in with them. I would take my problem
home and worry over it for a week, often finding that it would
in time resolve itself. But this did not always happen. I
remember well the occasion when Mr Godfrey, in a lecture on
properties of gases, introduced Dalton's Laws of Partial Pressure. One of these laws stated: 'If there are two or more gases
present in a given volume, then each gas fills the whole space
and exerts the pressure it would exert if it alone occupied the
space.'

My understanding of the nature of gases was hazy in the
extreme. 'Each gas fills the whole space,' I kept muttering to
myself, but I remained mystified by the concept. The following
week I decided to consult Mr Price, during the practical class.
Diffidently I presented my problem to him. To my discom-
fiture he burst out laughing. The unexpected depth of my
ignorance had presumably taken him by surprise. But then he
proceeded to enlighten me with a simple, straightforward
exposition. It was an exciting revelation. For the first time I
was able to picture a gas as a collection of tiny molecules
rushing about like hard little bubbles in what was almost
entirely empty space. There was plenty of room for more
molecules, of the same gas or of a different species, to occupy
the same volume. To give me a feeling for the smallness of
gas molecules, Mr Price pointed out that, while one can
visualize a kilometre length subdivided into a million millimetres—these dimensions being familiar in our every day
experience—it is much harder to think of a millimetre subdivided a million times; but such subdivisions are roughly
equivalent to the size of a molecule. There are an enormous
number of them in, say, a cubic centimetre of air—more than
a million million million; but they are nevertheless separated
from each other by many times their own size. Mr Price went
on to explain that the gas molecules are travelling about
randomly with enormous velocities—about a kilometre per
second—and are continuously bouncing off each other and off
the boundaries they meet; this is what produces gas pressure.

In just a few minutes, Mr Price had given me a whole new
insight into the physical nature of gases. I felt elated. What a

really exciting subject physics is, I reflected: full of surprises and fresh concepts to wonder at.

But the precariousness of my position at the Tech was suddenly brought home to me one day when, as I came into the practical class, Mr Price drew me aside and spoke quietly and seriously:

'An Inspector is supposed to be coming round shortly,' he said. 'We think it would be best if he didn't happen to notice you.'

He led me across to one side of the laboratory, where there was a long bench on which a tall set of shelves had been mounted. They were stocked with a variety of laboratory equipment and formed an effective screen. There was just room enough for me to perch on a stool behind the bench and to carry on with my work, completely cut off from the rest of the room. Mr Price nodded and smiled as he left me. I could hear the continuous murmur of voices, but otherwise could not tell what was going on beyond my hiding place.

After about half an hour Mr Price reappeared. 'The coast is clear,' he said, with evident relief. 'The Inspector has come and gone. You can come out now.' I thought of my first interview with Mr Godfrey, and remembered his saying that the "Authorities" might not approve of a schoolgirl at the Tech. I felt a surge of gratitude to him for being such a sport, and to both him and Mr Price for their personal help and the opportunities they were giving me. But after this episode I was apprehensive of being "found out", and I used to slink into the Physics Department for my classes, trying to look as inconspicuous as possible.

At one stage in the course, Mr Price's place as Demonstrator was taken for a while by an older man called Mr White. He was a kindly person, but, so I first thought, not as effective as Mr Price. However, he taught me an invaluable lesson which I have never forgotten. Because of school exams, I had had to miss some of the Tech sessions, and, in particular, I had not done any optics experiments. I asked Mr White if I could do one of these instead of the one scheduled for the day. He took me to a darkened lab equipped with a number of optical benches. Allocating one to me, he gave me brief instructions on how to measure the focal length of a convex lens and left me to get on with it.

I felt rather lost, on my own in this blacked-out room, but the procedure sounded straightforward enough. The lens was mounted at the centre of the bench, and at one end a pin, stuck in a cork and lit by a small lamp, was set up at a fixed distance from the lens. Looking along the bench I could see the inverted real image of the pin, sitting as it were in space. I had to determine its distance from the lens by sliding another pin along the bench, until it appeared to coincide with the image. The distance from the lens could then be measured. I found the procedure very tricky. My first attempts at judging, by parallax, when I had got the pin and image to coincide gave very dissimilar values for the distance. Apparently I had not yet got the hang of the technique. I kept on trying until I had achieved two readings which were reasonably consistent with each other. These I recorded in my lab notebook. The next measurement turned out to be quite different. I'd better ignore that, I thought, starting to get depressed. Carrying on, I got a lot of wild-looking results in between some values which agreed with the ones I had recorded. By now thoroughly demoralized, and concluding that I was no good at this kind of work when left on my own, I battled on until I had collected about six or eight sensible-looking measurements in my book. I calculated the average value and the error, worked out the focal length of the lens, and took my notebook along to Mr White.

He perused it for a minute or so in silence. Then, looking at me accusingly, he asked: 'Did you write down every measurement that you made?' Taken aback, I stammered something about there having been some which looked so erratic that I thought I should ignore them. 'Yes, I thought as much,' he said. 'I can tell, because your result looks too good, and the quoted error is much too small. This is not an accurate method of measurement, and a large scatter in the individual results is to be expected. You must understand that, in scientific experiments, some procedures are inherently less accurate than others.' He pointed out that, if all the observations are faithfully recorded, a reliable, if not very precise, average can be obtained. But if any attempt is made to fiddle selectively with the measurements, the result is worthless.

Blushing with shame, I felt as if I had been caught cheating. But Mr White gave me a kindly smile and said that what I had done was by no means uncommon amongst inexperienced students, and that I had probably learned my lesson for the future. He went on to talk about the paramount importance of integrity in scientific research: reputable scientists guard this principle jealously in their professional work. Misrepresentation of experimental observations is an unforgivable sin.

I have subsequently learned that inexperienced students frequently do fall into the same trap that ensnared me. There was a well known Cambridge physicist called Dr G F Searle, who ran the practical classes for undergraduates during the famous Rutherford era, and who went to extraordinary lengths to catch such students. Many stories are told of the methods he used. For example, he set up an experiment to measure *g*, the acceleration due to gravity, by timing the swing of a pendulum of a specified length. He had the bob at the end of the pendulum made of iron, and secretly hid a magnet under the bench on which the pendulum was mounted. In spite of this, so Searle used to say, a number of students contrived to fiddle their results until they arrived at the "correct" value for *g*. He then pounced on them ferociously.

I met Searle once, after he had retired. He was an impressive but very fierce-looking man, and I felt thankful that I had not had to start my physics studies under him—particularly as he had a reputation for being prejudiced against women students. He is said to have come up to an unfortunate female in one of his practical classes and to have said, aggressively: 'Are you wearing corsets? You cannot work in a physics lab, with its magnetic instruments, if you are wearing steel-boned corsets. Go and take them off.'

Fortunately I have not personally met such chauvinism. Mr Godfrey, Mr Price, and Mr White all treated me with special consideration and gave me every encouragement. My two-year course at the Sydney Tech was a great success, and, to my considerable surprise, I came top of the class in their final exam. Mr Godfrey was genuinely pleased that his venture in accepting me had turned out so well, but he told me, with a rueful smile, that the "Authorities" had finally found out about me, and a clear edict had been issued that no more schoolgirls should be admitted. So the door had been uniquely opened for me, and was firmly closed again after me.

During my period at the Tech the vicious grip of the Depression began to relax, and my father finally got a job. So, at last, my ambition to go to the University was transformed from a hopeful dream into a real possibility. The main requisite now seemed to be for me to do well enough in the

Leaving Certificate Examination to gain a bursary—a modest University grant, of which a small number were awarded each year, subject to a family means test.

We started to work out detailed plans. Since I was a year younger than the average age of my class, it seemed reasonable for me to put in an extra year in the sixth form at school, before taking the final exam. It was decided that I should sit the Leaving Certificate Exam twice: the first time just for the experience, taking my subjects at the pass level only. Then, in the second year, I would go all out for the best possible pass, taking the allowed maximum of six subjects with four at honours level.

At the first attempt I achieved my goal of A's in English, French, Latin, Maths I and II, Zoology, and, very gratifyingly, Physics. My best subjects at SCEGGS were English, French, and the two maths, so, for my final year, I chose these as my honours subjects, with physics and Latin at the pass level. The latter was belatedly substituted for zoology when my mother discovered that Latin was required for a lucrative scholarship awarded to the candidate with the maximum total marks in the Leaving Certificate Exam. This was obviously a long shot, but one which I might as well be in the running for. The only other scholarship open to me was that for the candidate topping the combined Maths Honours.

Since SCEGGS did not teach maths at honours level, my mother arranged for me to have private lessons from a man whose name, I well remember, was Percival Andersen. He was the head teacher at one of the big high schools for boys; I found his teaching excellent, and, in topics like differential calculus, quite enthralling. I had by then completed the Sydney Tech course in physics, but again my mother arranged a few private lessons for me, in order to keep the subject fresh, and to fill in gaps in the Tech syllabus. I even had a few sessions with our old friend, Scottie, to give me some practical hints for my French Honours. Not that my French teacher at school was not first-rate; likewise, my English one.

That was a really high-pressure year: I felt that there was so much at stake. I worked zealously, and my mother supported me with unremitting dedication. The exams, conducted during a particularly hot and sticky November, were an exacting ordeal. But finally it was all over, and I had only to wait until the first results appeared, one day during January, in the Sydney Morning Herald. I had got four first class honours, and two A's. The relief was tremendous. Already it

[margin note, handwritten: Math – Percival Andersen]

looked pretty certain that I would qualify for a bursary. The next day, the order of merit for the long list of first class honours in each subject was published: I had tied for second place in Maths I and II, with a boy named B F C Cooper; and I had come third in French and ninth in English. This was a very good overall performance, I knew, and my mother and I, and the School, were all highly delighted, though it was tantalizing to have missed, by such a narrow margin, the scholarship for coming top in maths.

Going through all the published results, I concluded that I must be in the running for the position of highest total marks. We had no idea how soon the scholarship results would be published, so my mother and I kept on scanning the newspaper eagerly each morning.

The papers were delivered from an open lorry which was driven slowly along the road while a man standing in the back rolled up each paper, gave it a deft twist, and flung it into the air so that it would land in the front garden of the house for which it was intended. Our dog, Billy, used to wait eagerly for ours and carry it triumphantly into the house for us. But occasionally the missile used to land in one of our tall bushes, where Billy couldn't reach it. This happened on what turned out to be *THE DAY*. I was awakened by Billy's howls of frustration. These turned to excited barking as my mother came to the rescue and gave the newspaper to him to carry inside. As I reached the table on which Mother had smoothed out the wrinkled pages, she suddenly exclaimed: 'They're here.' A second later she gave a tremendous shout of exultation.

'You've got it. You've got it!' she cried.

Bending forward, I read in wonderment:

THE UNIVERSITY OF SYDNEY
MATRICULATION EXAMINATION SCHOLARSHIPS

THE JAMES AITKEN SCHOLARSHIP, JOHN WEST MEDAL AND GRAHAME PRIZE MEDAL FOR GENERAL PROFICIENCY —

Freeman, Joan M (Sydney Church of England Girls' Grammar School).

And, as if that weren't enough, the announcement continued:

THE FAIRFAX PRIZE FOR GENERAL PROFICIENCY
AMONGST FEMALE CANDIDATES —

Freeman, Joan M (Sydney Church of England ... etc.

A wave of emotional delight swept over me. I hugged my
mother violently. What an unbelievable reward for my hard
work and her self-sacrifice. The euphoria we shared at that
moment was, I think, the greatest in all our joint experience.

Miss Wilkinson, summoned from her bath at my mother's
insistence over the telephone, received the news with incred-
ulity, followed by ecstatic pleasure. She had not known that
such an event was to be contemplated. 'Top of the State ...
Top of the State,' she kept repeating to me in a bemused voice
over the phone. Then, having adequately praised me, she paid
generous tribute to my mother's contribution.

So it was with an enormous boost to my self-confidence that
I was able to enter Sydney University, ready to stand up to,
and compete as necessary with, those young men who had
already gained Honours in physics and chemistry in the
Leaving Certificate. My objective was to study chemistry, *study*
physics, and maths, specializing finally in chemistry, since *chem*
everyone advised me that for a woman the subsequent opportunities in this field would be much greater than in physics.

Chapter 4

Evolution of a Physicist

The University of Sydney stands in a commanding position on a hill overlooking an old section of the City, about a mile beyond Railway Square. As I approached it, early in 1936, the honey-coloured sandstone of the main buildings, in their blend of Gothic and Tudor style architecture, glowed impressively in the morning sunshine. The Great Hall, the tall, richly ornamented clock tower, the cloistered quadrangle, and the spendid Fisher Library building with its steeply sloping, copper-clad roof, filled me with a sense of awe and privilege.

Beyond the main buildings, Science Road led to most of the Departments in Science and Engineering. Here I found the home of Chemistry, and, further on, the Electrical Engineering Department, with which I was later to become familiar. The Physics Building was set quite apart from the other science departments, in an attractive, open location overlooking tennis courts and a large playing field. This was my favourite building. It was very long (a tenth of a mile, I was told), with a low central section, framed symmetrically by twin towers and sturdy three-storied blocks. The creamy whiteness of the façade was relieved by large windows and dark columnar fir trees, planted at intervals along the front. Inside, the building was spacious; it was big enough to accommodate the Mathematics Department as well as the whole of Physics, so I got to know it well.

The first-year physics class was quite large, including students from other faculties, particularly engineering and medicine. There were a number of women, most of whom had learned little or no physics at school and found the going tough; I felt very grateful for my Sydney Tech training. Our lecturer was none other than Dr Edgar Booth, co-author of the

Figure 4.1 The author on matriculation day at Sydney University, 1936.

textbook *Physics: Fundamental Laws and Principles* that I had found so satisfactory at the Tech. He was a large, well-built man, with a military-style bearing and a direct, uncompromising personality. His lectures, eminently clear and logical, were relieved by occasional flashes of humour; he would introduce some quips and jokes, allowing the students a bout of shouting and stamping, after which he would demand and receive silent attention. I absorbed his course with enthusiasm. The practical classes were equally enjoyable, with a much greater range of experimental equipment than the Tech had been able to offer. I could see, though, that some of the male students were extremely capable and knowledgeable, and were a greater competitive challenge than I had ever met before. I would have to work hard to hold my own, I realised.

Before long I met Phyllis Nicol, Tutor in Physics at the residential Women's College and co-author of our textbook. In marked contrast to the extrovert character of Edgar Booth,

Figure 4.2 Phyllis Nicol. (Courtesy of the University Women's College)

she was a gentle, self-effacing person. Her straight brown hair was drawn tightly back from her pale face into an untidy bun, from which one or two wisps tended to escape, and did nothing to enhance her appearance. She usually wore a serious, slightly worried expression, but her full lips and thoughtful, wide-set eyes gave hint of a kind and generous nature, and a desire to please. I was impressed that such a gentle person had played such a pioneering role—she was only the second woman to graduate in physics at Sydney University. But, as I later learned, she had been frustrated in her efforts to pursue her career. I attended some of her tutorials at the Women's College and found her a very conscientious and dedicated, if somewhat uninspiring teacher; I suspect that her frustrations had subdued some of the fire and early enthusiasm she must have had for her subject. But she was very kind and helpful to me, and I was able to take to her some questions that I hadn't the courage to present to Dr Booth.

I had no real difficulties during my first-year physics, except for a hang-up about the subject of electricity. This had begun long ago when my mother refused to let me have an electric motor for my Meccano set, because she thought it was

electricity: prob

dangerous. Then, at the Tech, I had missed a significant section of the course on electricity because I had to be away for school exams. In my Leaving Certificate exam I had managed to dodge the topic. Now, at the University, I had to face up to my deficiency. I gradually overcame the problem, but to this day I tend to have a mental blockage when confronted with electric motors and generators. I suspect, though, that many physicists have their Achilles' heel.

In chemistry the pace seemed quite strenuous, since my initial knowledge of the subject was meagre. However, several women in the class were interested in pursuing chemistry as their major subject, and their company was helpful and stimulating to me. Physical chemistry I found particularly fascinating: the atomic nature of the elements, the structure of chemical compounds, and the satisfying logic of the periodic table, which classified the 92 elements into groups, according to their chemical properties. For the first time I grasped the fact that the number of electrons surrounding the positively charged atomic nucleus of a given element determined its chemical behaviour. The Head of the Department, Professor Fawsitt, was a rather intimidating person, tall, thin, and taciturn. But his inorganic chemistry lectures were very good. The practical classes in qualitative analysis were fun—a sort of detective game in which one had to identify samples of unknown chemical compounds by putting them through a series of tests. But quantitative analysis was tedious and required great care and patience.

Organic chemistry seemed to me very complicated and a rather messy subject, embracing all the compounds built upon the element carbon: from relatively simple kinds to the highly complex molecules associated with living matter. There were endless compounds to be classified and memorized. The subject was basically interesting, not to say fascinating in parts, but I disliked having to learn so much by rote. Fortunately we had an excellent teacher, Dr Frank Lions. He was a youngish man, big, hearty, with infectious enthusiasm and a lively sense of humour. He produced many entertaining dodges to help the recall of isolated facts. For example, the hydrocarbon gas *acetylene* is produced by the action of water on calcium carbide. 'Just remember,' said Frankie, 'the cat that ate the carbide had *a set o' lean* kittens.' I remember it to this day, though I have forgotten most of my organic chemistry.

In mathematics the going was hard for a different reason. From the start, a separate class was conducted for those students aiming to do Maths Honours for their B Sc. Following this course, it was possible to take the final Maths Honours papers at the same time as Maths III, after only three years, whereas in the other science subjects Honours required four years. Our honours maths course was intensive. The class was quite small—by the third year there were only eight of us, including one other woman and four electrical engineering students who had taken the option of spending an extra year to gain a B Sc before continuing with their engineering course. One of these was B F C Cooper, whose name was familiar to me since he tied with me for second place in the Leaving Certificate Maths Honours. He was tall, dark, and lanky, with a shy, hesitant manner, and was not easy to communicate with. When he spoke, it was in a series of reluctant, abbreviated bursts, always pertinent and frequently pungent. He was clearly very bright and at first I held him in some awe. I was to get to know Brian Cooper well, over the years, and my respect for his abilities was steadily reinforced. All the members of the maths class seemed earnest and hard working, and I felt that I had to be on my mettle to keep up with them.

We had a variety of maths lecturers. My favourite was R J (Dicky) Lyons, an experienced and effective teacher, with an inimitable style and an evident love of mathematics for its own sake. He guided us enthusiastically through the complexities of modern algebra, inspiring me with a lasting interest in the subject. During our third year, Dicky invited the whole class for a day out on his yacht, which he kept in Broken Bay, a lovely, unspoilt area about 20 miles north of Sydney. He proved to be as competent a yachtsman as he was a mathematician, and gave us a splendid day, including an island picnic, for which he brought all the provisions. I remember Dicky with great affection.

In second-year maths we met the Head of Department, Professor T G Room, recently appointed from Cambridge. He was a small, shy man of formal, very English manner, and gave the impression of having some difficulty in accommodating himself to the rougher Australian way of life, though he was eager to please. He was a very good mathematician, but I found his course on n-dimensional geometry memorably boring. On the other hand his lectures on the history of mathematics were fascinating.

Room tried hard to get through to his students. During our third year he invited us all to dinner at his elegant North Shore home. His wife, a cheerful, unassuming Australian whom he had recently married, did her best to put us at ease, but it was difficult to feel relaxed with the Prof. We were astonished when, as we sat down at the tastefully laid dinner table, the meal began to be served, not by a maid—that would have been unusual enough in an Australian home—but by a decorous *manservant*. I became so nervous that, while tackling a grilled chop, I shot some peas straight off my plate. Brian Cooper and the others, I could see, were equally ill at ease. However, we couldn't but feel a surge of affection for the Prof who was beaming hopefully at us through his thick glasses, in his well-meaning gesture of friendliness.

Overall, I enjoyed my maths at Sydney University very much, though I subsequently came to the conclusion that it had been too intensive, and somewhat inadequate in its coverage.

My first year was altogether a strenuous one. Not only did I work very hard on my University courses, feeling that I had much catching up to do, but I also had to give some support to my mother in a period fraught with difficulties. My beloved Grandpa died as a result of a street accident; then Grandma, becoming senile, had to be brought, unwillingly, to live with us; my father, put off by this situation, left home; and my mother still struggled with her school. By this time her oldest pupils, aged thirteen or fourteen, were contemplating the Intermediate Certificate exams in a year or two. She had to rely to some extent on my experience in the matter of syllabuses, textbooks, and specific problems, for the various subjects she was teaching. It was amazing to me that she could cope with English, French, Latin, history, geography, maths, and music, at that level, while at the same time keeping her younger pupils going. But I got great satisfaction out of being able to help a little; it was some compensation for the enormous effort my mother was making, for my sake.

The first-year exams were held in an atmosphere of pomp in the Great Hall. I remember being seated beneath the statue of a former dignitary who appeared to be staring at me with an expression of stony disdain. However, the results, when they appeared, were a tremendous surprise. I had gained High Distinction and first place in both Physics and Maths, Distinction in Chemistry, and a Credit in Zoology (my required fourth subject). Furthermore, I had won the George Allen Scholarship for Maths. It seemed unbelievable that I had done better than those very bright young men whom I regarded as so superior to me. Even in chemistry I had gained fifth place in a class of about ninety. It just showed what dedication and hard work could achieve.

In the second-year physics class I was the only woman, occupying, as it happened, a singularly isolated position during lectures. It was a University rule that attendance at lectures was compulsory. The method used for checking our presence was that of allocating a specific seat to each student; at the start of a class, an attendant came in and made a note of the absentees corresponding to the numbers of unoccupied seats. Women were always placed together in the front rows, and a gap was conventionally left between them and the male students. At the beginning of the year I found a seating list posted outside the physics lecture room. My place was in the front row on one of the aisles. Soon, a young man, after peering at the seat numbers, came and sat hesitantly beside me. The front row filled steadily except for the seat on the other side of my neighbour. I could see him becoming increasingly uneasy as the evidence for a mistake in his seat allocation mounted. Then our lecturer, Dr Booth, came in. After surveying his audience in silence for a minute, he stared quizzically at the man next to me.

'Officially you are not here,' he said. 'But if you feel strongly about sitting where you are rather than in your allotted place, I expect we can make an exception.' Loud guffaws arose from the audience. Blushing violently, the young man moved precipitately to the adjoining seat. He studiously avoided looking at me, let alone speaking to me, for the rest of the course.

We had other lecturers besides Dr Booth. There was Professor O U Vonwiller, the Head of the Department. He was a pleasant man, but quiet and reserved. I never really got to know him, but I appreciated his lectures, particularly those concerned with the nature and properties of light. It was exciting to learn about light waves being a form of electromagnetic radiation, like radio waves but with wavelengths less than a millionth of a metre. Sometimes light behaved like separate little particles or bundles of energy, called quanta, being emitted from individual atoms when the orbital electrons dropped from higher-energy to lower-energy states.

Another of our lecturers was Dr G H Briggs, who was to make a significant impact on me later in the year. He was a tall, dignified, good-looking man, gentle, but quietly self-confident. His lectures were delivered in a slow, somewhat hesitant and expressionless voice, but with a clarity and simple logic which arrested the attention of the class. I remember particularly his course on the general properties of matter.

The Physics II practical classes were also interesting and varied, and those relating to electricity helped to dispel my sense of inferiority about this subject. But the practical classes in chemistry—particularly in organic chemistry—seemed less attractive and often tedious. Many of the procedures were empirical, rather like those in an advanced cookery book, I thought. In one experiment we had to produce a permanent dye in a sample piece of flannel. 'Stir thoroughly,' the instructions said, at one stage. I agitated my glass stirring rod vigorously, poking at the flannel which was floating uncooperatively in my beaker. Suddenly the rod went straight through the bottom of the beaker, and the potent, evil-smelling liquid poured down over my stockings, staining them indelibly. 'I hate organic chemistry,' l muttered angrily to myself, thinking partly of the not negligible cost of lisle stockings, but mainly of the indignity of the situation. I wondered why I was intending to go on with chemistry for my degree, just because everyone said I should. Why couldn't I become a physicist instead? I decided to go and talk to Phyllis Nicol about it.

But Phyllis was not very encouraging. 'It's difficult enough for a man to get a job as a physicist,' she said. 'But for a woman the possibilities are very limited indeed.' She described her own struggles to get a foothold in the profession. After gaining her B Sc in physics, with a scholarship, in 1926, and an M Sc, in 1927, she achieved only a temporary position, as a Demonstrator in the Physics Department. This position was renewed annually, but in 1932 was terminated, in accordance with a faculty decision to limit the duration of temporary appointments. She had had no opportunity to do original research after completing her M Sc, and had been able to publish only one paper. In 1933 she became Tutor in Physics at the Women's College, but this was not a very rewarding job as she had only first-year students, mainly medical, who had no real interest in physics, being bent on scraping through their compulsory year in the subject as painlessly as possible. As for the possibility of getting a job as a physicist in industry, this, for a woman, seemed to Phyllis unlikely. The opportunities in chemistry looked much brighter. She felt it would be a safer line for me to follow.

Chastened, I went away to brood over the situation. In retrospect I can see that, given the pre-war conditions in which Phyllis Nicol found herself, her gloomy prognostication may well have been justified. Had it not been for the onset of World War II, I might have found myself, on graduation as a physicist, in an unenviable position. But in the event developments and applications in the physical sciences were given a tremendous boost, and many exciting jobs became available both during and after the war. Undoubtedly the timing of my arrival at the University was more fortunate than it had been for Phyllis.

In spite of Phyllis' discouragement I still felt strongly drawn to physics, since I was enjoying the subject so much. Then two events helped to strengthen my attitude.

First, one day, a new sound was heard echoing through the long corridor of the Physics Department: an unfamiliar voice, accompanied by penetrating, high-pitched bursts of excited laughter. It came from Professor V A Bailey, Professor of Experimental Physics in the Department, who had just returned from a visit to Europe, where he had been carrying out an original radio experiment. He was a distinctive, vigorous little man, dark-complexioned, with a small moustache and bright brown eyes beaming intensely through small metal-rimmed glasses. He seemed to bounce rather than walk, exuding an air of enormous enthusiasm. He enjoyed talking, usually about some aspect of physics: some phenomenon for which he had discovered an original interpretation, or some new calculation he had just made. His expositions, often delivered as he bounced along the corridor with a captive listener, were punctuated with cackles of delight at the new marvel that he had uncovered, or simply at his own ingenuity. Though quite egotistical, the naive, almost childlike quality of his outlook, and his unalloyed pleasure in studying the natural world, were so disarming that one could not but warm to him. He was very absent-minded, and used to leave behind him a trail of coats, hats, brief cases, and so on. His wife had become accustomed to finding out his movements and gathering up his scattered possessions.

He proved to be one of the world's worst lecturers, with an irritating habit of clearing his throat at frequent intervals as he sought to collect his racing thoughts into logical statements. His heart was clearly in research rather than in teaching; nevertheless his enthusiasm was very infectious. He

Figure 4.3 Professor V A Bailey. (Courtesy of the Sydney University Library)

had soon recounted to everyone in the Department, including his students, the nature of the radio experiment that he had been carrying out overseas. I describe here some details of the memorable lecture he gave us, because it was to prove of particular significance for me.

Bailey's story was about a phenomenon in radio broadcasting first reported in 1933 and referred to as the "Luxembourg effect". The reception at Eindhoven in Holland of a medium-wave programme broadcast from a station in Switzerland was found to be disturbed by a background of the long-wave programme transmitted from Radio Luxembourg, which was about half-way between the two. Many other examples of the effect were soon observed. In 1934 Bailey, together with Dr D F Martyn, a physicist at the Australian Radio Research Board, put forward a successful theory to explain this phenomenon in terms of effects produced by the radiowaves on electrons in the ionosphere (the rarefied layer of free electrons and ionized gases at the top of the Earth's atmosphere, which reflects radiowaves and makes long-distance radio transmission possible). Bailey described how, in the previous year (1936), he had found a small inaccuracy in the theory. On revising it he noticed that a pronounced resonance effect was predicted to occur when the frequency of the long-wave transmitter was at or near the natural frequency of oscillation of the ionospheric electrons in the Earth's magnetic field.

With rising excitement Bailey then launched into a dramatic account of how he had gone to Europe earlier in the year to organize experimental tests of his resonance theory. It was a remarkable story. He had succeeded in persuading eleven broadcasting stations in seven different countries — England, Ireland, France, Belgium, Holland, Luxembourg and Switzerland — to cooperate by transmitting special prearranged signals on various nights during a two-week period between five minutes and forty-five minutes past midnight. I could imagine him, in his inimitable fashion, tackling and overwhelming a high-ranking member of the British Broadcasting Corporation, and working from there. He had also organized numerous volunteer observers to listen in on radio receivers in various parts of Britain, Ireland, France, and Belgium and to record what they heard during the experimental period. I pictured him bouncing up and down with exuberant enthusiasm as he made his own observations. Finally Bailey showed us the results, which strikingly confirmed his resonance theory, and, with a triumphant flourish, concluded his lecture, his bright beady eyes sparkling with self-satisfaction.

I felt exhilarated. Physics research might not always be as dramatic and original as this, but it was evidently an exciting activity. Could I aspire to becoming a research physicist? It would not be easy, I reflected, recalling Phyllis Nicol's cautionary comments about the lack of opportunities for women. I was in fact lucky to be able to study physics at all. The Engineering Department, I knew, had never had a female student. I had heard the possibly apocryphal story of a woman who insisted on being accepted, having pointed out that there was no written rule forbidding the admission of a woman. But the Department retaliated by setting her tasks like dismantling a large, oily engine, and other unpleasant or physically difficult pieces of work. In the end she was forced to give up. With prejudices of this nature, how would I fare in seeking a job in competition with an obviously competent man, the like of Brian Cooper, for instance?

Later in the year, after I had got to know Bailey a little, and he me, I consulted him about the possibility of my pursuing physics, rather than playing safe with chemistry. Beaming characteristically, he assured me that if I wanted to go on with physics I should do so. He was confident that I would do well, and would find a suitable outlet for myself. Although I was aware that Bailey's judgement might not be very balanced, his encouragement was pretty irresistible, and I felt more strongly inclined than ever towards physics.

Then another event occurred which had the effect of drawing me further in this direction. One day, towards the end of October, we were waiting for Dr Briggs' usual physics lecture when he entered the room wearing an unusually serious expression. He stood for a moment in silence and then, in a voice charged with emotion, he announced:

'I have just heard the news that Lord Rutherford is dead. Instead of my scheduled lecture I am going to talk about this remarkable man, one of the greatest experimental physicists of all time, whom I have been privileged to know and work with in the famous Cavendish Laboratory at Cambridge.'

With an eloquence of which I had not thought him capable, Briggs conjured up for us a vivid picture of "the Prof": a big man, with a boisterous personality and an intense enthusiasm for physics, his booming voice and remarkable presence pervading the whole of the Cavendish. His intellect, his scientific abilities, and his capacity for inspiring people were astounding, Briggs said. We heard about some of Rutherford's brilliant achievements, such as the interpretation of radioactivity in terms of the emission of alpha particles (identified with helium nuclei) and beta particles (equivalent to electrons), and the discovery that the tiny nucleus at the centre of an atom was made up of positively charged protons and their neutral equivalent, neutrons. Briggs also mentioned how Cockcroft and Walton produced nuclear disintegrations with a beam of accelerated protons, and I recalled with a thrill the newspaper account that I had seen as a schoolgirl.

Briggs then described his own research work at the Cavendish. His whole talk made a deep impression on me. This was not simply because of the stirring picture he painted of the great man and the Cavendish atmosphere, but also because of the sudden revelation that Briggs himself—this rather dry lecturer—was imbued with an intense interest in physics and a fascination for research, particularly in nuclear physics, which evidently continued to be a fulfilling activity for him.

This event reinforced my desire to pursue my studies in physics and clinched my resolve to become a physicist. With an easier mind I threw myself into the preparations for my second-year exams. The results were highly satisfactory: first place, with High Distinction, in both Physics and Maths, a

Figure 4.4 Dr G H Briggs. (Courtesy of CSIRO Archives)

position I was to maintain in the following year; a Credit for Chemistry; and the Slade Prize for Practical Physics. The latter was particularly gratifying as I was already conscious of the fact that experimental physics appealed to me even more than theoretical physics.

For my mother too, this was a very successful year. Both her Neuchalet girls taking the Intermediate Exam passed— one very creditably, the other with a mediocre result. That the latter passed at all was a triumph, since her parents had brought her to Neuchalet at the beginning of the year after her previous school had refused to prepare her for the exam, saying that she would be sure to fail.

In my third year, at the end of his last lecture of the first term, Bailey made a surprise announcement which proved to be important for me.

'I suggest that during the vacation you write an essay on

the magnetron,' he said casually to the class. He explained
that this was a new type of vacuum valve for the generation
of ultra-high frequency oscillations, and warned us that we
would not find it mentioned in textbooks. 'You will have to
look up the original papers about it, in recent issues of
scientific journals,' he said. 'It will be good experience for you
to learn how to use the scientific literature.' I discovered some
time later that it was usually not until the post-graduate level
that students were expected to go through this kind of
exercise.

We were all taken aback. The vacation period was precious,
and none of us fancied having to spend a significant part of it
in this way. However I thought I had better try to satisfy the
Professor, so I decided to start on the project straight away.

But how to go about it? I enquired first at the Main
Library. They suggested I try the Electrical Engineering
Department. There I found the office of the Secretary to
Professor J P V Madsen.

'I'm Doris Wood,' she said pleasantly, when I had introduced
myself and explained my problem. 'I think we can help you.'
She led me to a large room lined from floor to ceiling with
shelves carrying massive volumes of uniform appearance. 'I
expect the journals you'll need to refer to are all here,' she
said. 'I'll just go and ask Professor Madsen if he agrees to your
using our library.'

While she was gone I contemplated with dismay the overpowering array of books. Peering tentatively at the spine of a
nearby volume, I read: "*Journal of the Institution of Electrical
Engineers*, vol 72, 1933." It must run to more than a thousand
pages, I reckoned, and there seemed to be a hundred or more
such books lining the walls.

Doris was soon back, saying that the Professor, whom she
clearly revered, was quite happy for me to make use of the
library. Then she noticed my expression. 'Are you not familiar
with using scientific journals?' she asked. 'I've never seen
anything like this before in my life,' I replied. She laughed in
a kindly way. 'You'll soon get used to it. I suggest you start
with the Science Abstracts. These are published monthly and
provide titles, abstracts, and references for papers in all the
journals, under various subject headings.' She pointed out
their location on the shelves. 'See how you get on, and let me
know if you need some help.'

I took a recent unbound issue and began to study it. There
seemed to be about a dozen subject headings, starting with

"Prime Movers and Kindred Plant." What ever was that? I found that it was about engines. There followed a number of similarly unpromising titles, but finally I came to "Radio Communication." This seemed a possible category for me, so I read through the several pages of its entries. There were papers on a variety of subjects, but nothing remotely connected with magnetrons.

After lunch with Doris, I waded through a dozen or more issues of Science Abstracts with similarly fruitless results. I was becoming mentally tired and decidedly depressed. Then suddenly I saw an entry with a mention of magnetrons. I located the journal and found the paper, which proved to be quite informative, and also gave references to earlier papers on the subject. I was in business at last!

The magnetron, I discovered, was a valve consisting of a cylindrical anode, longitudinally split into two halves, surrounding a straight wire filament, along which a magnetic field was applied. Electrons emitted from the filament followed circular or spiral paths because of the magnetic field, and under some circumstances oscillations could be produced at ultra-high frequency, corresponding to wavelengths of less than a metre (nowadays known as the microwave band). Most of the papers I looked at were concerned with theories of how the oscillatory process occurred. They seemed complicated and confusing. After spending several days trying to sort them out, I began to worry about the time all this was taking. I hadn't seen a sign of any of the rest of the class; evidently they weren't taking the assignment too seriously. I wondered if I should draw the line and start writing. But my nature was to persevere until I had mastered the subject, so I plodded on and finally produced what I felt to be a satisfactory essay, of some thirty odd pages plus a substantial bibliography.

At the beginning of the next term, when we came to hand in our essays, I discovered that the others had indeed left it to the last minute to pursue the exercise, and their offerings were only ten pages or less in length. Had I been over-conscientious? Well, it seemed not when Bailey in due course returned our work to us. After the class, he singled me out and told me that he was delighted with my paper. Beaming typically he said that he had found it informative and helpful, as he was intending to acquire some commercially made magnetrons for an experiment he had in mind. This seemed an adequate reward for the big effort I had put in, and I felt gratified. But it was not until the following year that I discovered just how significant that episode was for me.

During the year I attended a course of lectures which I found particularly fascinating. They were given by Dr D F Martyn, Bailey's co-author in the original theory of the Luxembourg effect. Martyn was a member of the Radio Research Board, based in the Electrical Engineering Department under Professor Madsen, and was, I discovered, playing a prominent part in the front-line research that they were doing on the ionosphere. An interesting, unusual-looking man, with thinning hair and a large, pale face, he proved to be an excellent lecturer. He talked about the different levels of the earth's atmosphere: the troposphere, the ozone layer, the stratosphere, and, particularly, the ionosphere. He described how the complex layers of ionized gas in the latter were being studied by the reflection from them of very short duration pulses of high-frequency radio waves, and he discussed the theory that the ionosphere was formed by charged particles emitted from the sun. Then he pointed out that a visible manifestation of these solar particles occurred in the phenomenon of the aurora—bright streamers of light sometimes appearing in the night sky at high latitudes, near the earth's magnetic poles. Finally he explained in simple terms some of the features of meteorology: Coriolis forces, highs and lows. I still tend to think of Dr Martyn when I see the weather maps on television. It was a memorable set of lectures he gave, and the respect I then developed for his ability as a research scientist and as a lecturer was not diminished by subsequent events, which I shall recount later.

This was my final (Honours) year for Maths, and, although I was intending to spend another year doing Physics Honours, I would be technically qualified for a B Sc by the end of the year. My mother could see that her great mission—to ensure that I had a University degree—would soon be fulfilled. Her school and her present burdensome way of life had played their essential part; she now yearned for a change and an easier existence. On the strength of the public examination results that she had achieved, and the reference that Dorothy Wilkinson, the SCEGGS Headmistress, was prepared to provide, she applied for the job of Mathematics Mistress at the Presbyterian Ladies College. This was a major private girls' school at Pymble, a suburb up the North Shore line, about twelve miles from the City Centre. She was offered the post, and decided to accept it, to close down Neuchalet, to abandon the Vaucluse house—since there was no prospect of the mortgage ever being paid off—and to rent a house in the area of Pymble. I was delighted with her plan, as I had been feeling very unhappy about her continuing to lead such a hard life while I was enjoying myself so much.

In January, 1939, we found a very pleasant house to rent in Turramurra, the next suburb along the line from Pymble. It was within walking distance of the station, on Kissing Point Road, which dropped down the side of a hill in a steep S bend, and then, not far beyond our house, petered out, giving way to a rough track which led enticingly for several miles through virgin bush to a spot on the Lane Cove River known as Kissing Point.

The generously-proportioned house was virtually surrounded by open land with its natural scrub and tall gum trees. From the back verandah there was an extensive view over an undulating sea of gum trees towards the City, whose distant lights could be seen faintly twinkling at night. The garden was visited by many birds, such as kookaburras, magpies, wagtails, butcher birds and wrens. We appreciated very much the peaceful, countrified atmosphere, and even Billy, by then quite an age for a little dog, seemed rejuvenated as he ran around enthusiastically, exploring the new environment. Above all, the relative ease of her new job gave my mother a new sense of freedom and the opportunity to enjoy a little leisure, which pleased me greatly.

With a new chapter opening up in my home life, my Maths Honours behind me, and the Deas–Thompson Scholarship for Physics III, I started on Physics IV with a light heart. There were only six of us now, Brian Cooper and the other engineering students having gone off to complete their engineering course. We found ourselves thrown on our own resources more than in previous years. Bailey ahemed his way briefly and somewhat incomprehensibly through special and general relativity (the latter pretty useless to me, I thought), but gave us a solid course on electricity in gases, which was to prove of value to me in the future. Unfortunately we got no lectures in nuclear physics, as Dr Briggs had left to become Head of the Physics Division of the new National Standards Laboratory.

We had one very frustrating course, on quantum mechanics, the fundamental mathematical theory on which much of modern physics is now based. Our lecturer, only just appointed to the Department, was Dr R E B Makinson, a tall, rather hesitant, fair-haired young man, fresh from Cambridge, where he had been working for his Ph D degree in theoretical physics. He was clearly immersed to great depth in his subject, and was unable to visualize the level of understanding of students completely new to it. He tended to talk in generalities rather than in crisp mathematical statements, and had us floundering from the start. We could not even formulate questions to ask him. When he introduced something called the Uncertainty Principle, he raised in us only a cynical smile. I realise now that his trouble was a total lack of lecturing experience, for he subsequently became a good and popular teacher. But I was very much shaken at the time by this apparent breakdown of my ability to grasp and learn something new. It was to take some fifteen years before my self-confidence was fully restored as far as quantum mechanics was concerned. I then had a brief but excellent course of lectures at Harwell from Brian Flowers, a very bright young theoretician with a tremendous teaching flair. I learned from him that the Uncertainty Principle, brilliantly conceived by the German physicist Heisenberg, is a vital cornerstone of modern physics and leads to physical predictions of great accuracy.

As I got to know Dick Makinson better, I found him basically very likeable—kind and gentle, though unsociable and generally uncommunicative. This may have been mainly through shyness; occasionally he would become eloquent, for instance when expressing his surprisingly strong left-wing political views. Towards the end of the year he astonished the class by suddenly inviting all six of us at short notice to a party. We turned up diffidently, wondering what to expect. Without ceremony, he introduced us to Rachel, whom he had just married at a Registry Office, a few hours after her arrival from Cambridge. They had met while Dick was in Cambridge and she was studying for her B A in physics. She had left for Australia as soon as she got her degree. Through Dick's instigation she had been given a temporary appointment as a Research Assistant in the Physics Department. She was a large, to me rather intimidating person, remarkably self-possessed for someone who had just arrived in a strange country. She was very articulate, with quick, decisive movements, and gave the impression of being highly intelligent, knowledgeable and capable. Her directness and friendliness were attractive, and my affection, as well as my respect for her were to grow.

I soon experienced an example of Rachel's drive and determination. Shortly after the party, at around midnight, I was aroused from sound sleep by the insistent ringing of the telephone. Answering it reluctantly, I heard a very English, peremptory voice:

'Joan, this is Rachel Makinson. Do you have a sewing-machine?' I tried to summon my faculties.

'My mother has one which I use occasionally,' I replied hesitantly, thinking of the old treadle machine we'd had for years. I hoped she wasn't going to ask to borrow it.

'I have just acquired a second-hand one and I can't get it to work. I thought you might be able to advise me.' She launched into a highly technical discourse, bombarding me with words like clutch, gear, shaft and cam, shuttle and bobbin, take-up lever, and tension disc.

My mind boggled. 'I've only just wakened up,' I said weakly. 'I really couldn't follow what you were saying.'

Speaking more slowly, she repeated more or less what she had said before. How was I to put her off? Then I thought of a suitable response.

'I'd have to go and look at our machine to understand your problem, and I can't do that now without waking my mother.'

'Oh, alright. Sorry to wake you,' she said matter-of-factly. The implication seemed to be that Australians must go to bed ridiculously early.

The next day, coming into the Physics Department, I wondered apprehensively what the sequel was going to be. But Rachel was full of smiles, announcing triumphantly that she'd managed to work it out for herself.

It was not surprising to me that Rachel Makinson, as well as producing two very bright children, succeeded in rising high in her profession, ultimately becoming Chief Research Scientist and, for a time, Assistant Chief of the CSIRO Division of Textile Physics. She is the only woman to have reached these levels in the CSIRO (Commonwealth Scientific and Industrial Research Organization, the title adopted by CSIR in 1949).

For me, the most absorbing feature of my honours year was the practical work. First, we learned interesting techniques like glass-blowing, metal work, and soldering. We also had a special course on mechanical design and drawing, held in the Department of Mechanical Engineering. It was given by an amiable, patient, and very experienced teacher called Geordie Sutherland. From that brief course I acquired the essential techniques for producing simple workshop sketches—an invaluable asset to me in the years to come. I remember the

Figure 4.5 Some of the lecturers and students at one of the special radio courses run for the Services by Professor Bailey at the Physics Department during World War II. Standing, from left to right: first, Rachel Makinson; third, V A Bailey; fifth, Mrs Joyce Bailey. (Courtesy of Mrs K R Makinson)

surprise of the foreman of the local workshop at Harwell when I provided him with "proper drawings—not like some male scientists he could name."

The most important part of our practical work was the special research assignment which each of us was expected to carry out and write up. Bailey chose to give me the experiment he was directly interested in—using a magnetron oscillator. He had recently bought some magnetrons from the Philips Electrical Company and he wanted them to be set up in an oscillatory circuit to create an ultra-high frequency electric field. His objective was to use this to produce a gaseous discharge at low pressure in the presence of a magnetic field, to simulate at the higher frequency the resonance effect that he had been investigating in the ionosphere. I could see now why he had set us the task of looking up references on magnetrons the previous year; why he was particularly pleased with the paper and bibliography that I had produced; and why I was now chosen to work on his pet experiment. My slog over that vacation period was now reaping its reward.

A pair of split-anode magnetrons was used, operating at a wavelength of 60 centimetres. Fortunately for me, there was a senior Research Assistant called Clive Davis and a competent Technician, Bob Wilson, who undertook much of the design work involved in making the required apparatus, and I was able to apply myself to the testing and subsequent experimenting. On the outbreak of World War II, Davis promptly went off to join the RAAF, but by then I was capable of carrying on—with some advice from Bailey and support from Rachel Makinson—and had the great satisfaction of seeing the resonance effect demonstrated qualitatively before the end of the year.

Bailey's enthusiasm for the experiment continued unabated. That his interest in it extended beyond the immediate study of ultra-high frequency electrical discharges in gases was revealed when he gave a colloquium in the Department with the title: "Generation of Auroras by Means of Radio Waves". He proposed that, given sufficient power from a radiowave transmitter operating at the resonance frequency, a visible glow discharge in the ionosphere could be produced. He called this effect an "artificial aurora". Furthermore, he made the startling suggestion that this glow could be made bright enough to illuminate the streets of greater Sydney on clear nights with an intensity equal to that of the full moon overhead. The transmitter would need to radiate about a million kilowatts; this, he declared, would be an effective alternative to present methods of street lighting. I learned later that he had filed a patent application for this suggestion.

The audience seemed mostly to receive Bailey's bizarre

proposal with tolerant amusement, but Dr Martyn, who had been sitting aloof at the back of the lecture theatre, stood up at the end of the talk and made some biting comments. He objected particularly to Bailey's use of the term "artificial aurora", declaring that it was insupportable because an aurora was a phenomenon caused by the action of charged particles emitted from the sun. This seemed to me a case of hair-splitting, but it confirmed what I had already suspected: that Bailey and Martyn had fallen out since their original collaboration, and there was now bitter enmity between them. I was to have personal experience of this before long.

Bailey was ambitious to obtain financial support for mounting a pilot experiment to test out his scheme for a glow discharge in the ionosphere. He was therefore very keen for the laboratory experiment to go ahead, to provide quantitative results which would help to verify the details of his calculations; moreover the experiment was a tangible demonstration for the benefit of potential providers of funds. He proposed that I continue with the work during the next couple of years, using it as the basis for an MSc thesis. I would be financially supported by a grant of some kind, he assured me. The offer sounded attractive, and I hesitated only because by this time Australia was very much committed to participation in the war, and already a number of my contemporaries were moving into war work. However, Bailey told me that he would soon be starting up some war-related research in the Department, and that I could contribute to this for part of my time. So, with my final honours exams completed, and a resulting BSc with a double first plus a Commonwealth Research Scholarship, I launched happily into 1940, as an MSc student.

Having now a regular income from my scholarship, I decided to open a proper bank account.

'Name and address?' asked the young man behind the counter, when I explained what I wanted. He wrote them down laboriously on the form he had produced.

'Occupation?' he then asked.

'Physicist,' I replied proudly.

'How do you spell that?' he asked.

I was taken aback at the unexpected question. 'P . . . h . . .,' I began, hesitantly. How did one spell it? How many y's, i's, s's, c's? If I could try writing it down, I might be able to work it out, I thought. But the young man, staring blankly at me, had me unnerved. 'Oh, just put down "Scientist",' I said in desperation. That taught me not to be too cocky about my newly won status.

My experiment continued to be interesting and absorbing. In due course I was able to measure the resonance effect in various gases and to provide satisfactory verification of Bailey's theory. Bailey followed the work closely and cackled with pleasure when the results started to appear. He induced many visitors to come and see the equipment in operation, hoping that he could persuade one of them to support his artificial aurora scheme. They were all pleased to see our glow discharge appear as the magnetic field was swept through the resonance value, but then they invariably just nodded and went away. The most distinguished visitor we had was Professor R A Millikan, the American Nobel prizewinner famous for having determined the charge on the electron. He was a large, jovial man with a shock of white hair, and followed attentively Bailey's discourse on the experiment. But when Bailey went on to say that he needed financial support to carry out a test in the ionosphere, Millikan slowly shook his head. 'If you can think of a way of writing "Coca-Cola" in the sky, you might get some money,' he said, laughing heartily. But Bailey was not amused.

He never did get the full financial backing he wanted, though, some years later, a low-power pilot experiment was run with some success. The trouble with Bailey was that his behaviour, and the ideas he put forward, were sometimes so eccentric that many people refused ever to take him seriously. But I developed a great affection for V A, as he was usually called. His enthusiasm for physics, his sharp intellect, his child-like exuberance, and his personal encouragement, were an inspiration to me. Mr Godfrey and Mr Price at the Sydney Tech had put me on the first rung of the ladder to a physics career. It was Bailey who set me up on the next rung.

One of Bailey's idiosyncrasies was a special interest in numbers and the history of individuals reported to have been capable of prodigious mental calculations. During 1940 he gave a public lecture on the subject, the climax of his talk being a demonstration of how he could teach people to perform feats of a similar nature. For this purpose he had more or less compelled his wife and myself to act as "guinea pigs": standing up in turn and multiplying mentally two six-figure numbers which the audience had provided. But the performance was in fact a deception which he did not explain to his listeners. He had devised a method for such multiplications which did not require more than a reasonable ability in mental arithmetic. In an atmosphere of conspiratorial secrecy he had, a few days in advance of the lecture, revealed the procedure to us, and put us through a practice session. It was an amusing technique, most easily described for the simple case of two three-figure numbers. Take, for example, the multiplication of 746 by 532. We were allowed to write down just the two numbers, one on each of two cards. The trick was to write the second number *in reverse*. Thus at the top of one card we had 746, and at the top of the other, 235. The second card was slid along to the left relative to the first. The procedure was as follows:

With the dispositions
    746
        235 calculate $(2 \times 6) = 12$; carry 1 and call out 1st digit 2,
    746
        235 calculate $1 + (2 \times 4) + (3 \times 6) = 27$; carry 2, call out 7,
    746
    235 calculate $2 + (2 \times 7) + (3 \times 4) + (5 \times 6) = 58$; carry 5, call 8,
    746
    235 calculate $5 + (3 \times 7) + (5 \times 4) = 46$; carry 4, call 6,
    746
    235 calculate $4 + (5 \times 7) = 39$; call out 9, 3.

Bailey wrote up the digits from right to left on the blackboard as they were called out, giving the final answer which, in the example above was 396872. With the larger numbers that we actually had to work with, the mental arithmetic required more concentration because two-figure numbers had to be carried forward, but we found this to be within our capacity. To Bailey's evident satisfaction, the audience seemed duly impressed, but I felt uncomfortable with the deception: there was nothing prodigious about what Joyce Bailey and I had done.

During term time the New South Wales Division of the Institute of Physics met periodically in the Physics Department, a talk being given by a staff member or a visitor on a subject of topical interest. One day Bailey came bouncing up to me suggesting that I give a talk at the November meeting on the subject of klystron oscillators. The klystron, he explained, was a very new type of ultra-high frequency device which I could read up about in recently published journals. He was sure that the subject would interest a number of people. I protested that I could hardly stand up and lecture for the best part of an hour on something that I had never even seen. No one in Australia had seen one, he assured me, and by the time I had studied the subject as thoroughly as I had the magnetron, I would know more about it than anyone in the audience. Once Bailey had made up his mind about something it was very difficult to resist him, so in the end I agreed to do it.

Once again, in some trepidation, I resorted to the Electrical Engineering Department Library. Doris Wood greeted me warmly, and even the distinguished Professor Madsen nodded to me in a friendly way. It was much easier this time to find my way around the scientific literature, and I soon discovered that the klystron, which had been made for the first time during the previous year (1939), in the USA, was much simpler to understand than the magnetron. The principle of velocity modulation, on which the idea of the klystron was based, was interesting and ingenious and I found ample material from which to prepare a reasonable talk. I went over it meticulously before the event.

When the time arrived, I entered the lecture room in a state of nervous tension reminiscent of the sensation before an exam. This was not improved when I saw what seemed to be a huge audience, with many strange faces, including, I learned afterwards, men from AWA (Amalgamated Wireless of Australasia) and other engineering firms. Bailey gave a few words of introduction, and then I found myself facing the expectant gathering. With a sinking feeling I wondered how I was going to maintain their attention for an hour. I began falteringly, though without betraying my extreme nervousness, I was told. Then, plunging into my subject, I concentrated on the story, trying to convey the substance and special interest of this new invention. Sensing that the audience was with me, I felt my apprehensions fall away, and my talk flowed on with increasing confidence, finally reaching what seemed to me an adequately convincing conclusion. A gratifying round of applause followed, and Bailey was beaming with satisfaction. A surge of relief swept through me.

But then the dreaded moment came when the meeting was opened to questions and discussion. Would this be my undoing? There were a couple of simple questions which I was able to answer adequately. Then a serious-looking man stood up and asked a practical question about a klystron under operating conditions. I hardly understood what he was asking, and I certainly hadn't a clue how to reply. I stood in consternation, speechless and embarrassed. Then, slowly and hesitantly, a tall, young-looking, loose-limbed man, with an open, boyish expression, stood up near the back of the lecture theatre and said that he could perhaps answer the question. Apologising for the fact that his experience of klystrons in operation was very limited, he proceeded to interpret in simple terms the point that the questioner had raised. The man seemed satisfied. There were no further questions and the meeting was closed. Weak with gratitude, I approached my rescuer to thank him for saving the situation for me. 'Evidently you, rather than I should have been giving this talk,' I said to him. 'Oh, no,' he replied, his face breaking into a warm, gentle smile. 'I just happened to see a klystron operating recently at EMI (Electrical and Musical Industries) in England, where I was working in a completely different field. You know much more about klystrons than I do, and I enjoyed your lecture.' I could have hugged him.

'Who is that man?' I asked Brian Cooper a few minutes later.

'He's Dr Pawsey,' replied Brian in his matter-of-fact way. 'One of the key men in that new outfit just up the road.' He gave a characteristic slight jerk of the head in a vaguely easterly direction.

Chapter 5

The Radar Days

"That new outfit just up the road" referred to a large, light-coloured brick building recently erected just beyond the top of the road which ran past the Physics Building. It was part of a big new block built for the Council for Scientific and Industrial Research (CSIR) to house a National Standards Laboratory. Above its modest, closely guarded entrance, was carved the name *Radiophysics Laboratory*.

From Brian Cooper and his fellow engineering students I gathered that the Lab, headed by Dr D F Martyn, was engaged in some highly secret and important radio work connected with defence. They were being urged by Professor Madsen to go into this work when they had completed their engineering degree course. Several of the brightest graduates from the previous year had already joined the laboratory. All this sounded very intriguing to me. As 1941 got under way, I was beginning to feel restless in my present job. I had accumulated enough data on the ultra-high frequency resonance experiment to be able to start writing my M Sc thesis, and the odd bits of war-related research that Bailey had given Rachel Makinson and myself to do were not very satisfying or productive. Later Bailey was to make a substantial contribution to the war effort, organizing and directing crash courses in radio and radiophysics for the Australian Armed Forces, but his initial ideas were hardly practical. I felt that I should be more directly involved in war work.

Then an advertisement appeared inviting applications for appointment to the staff of the CSIR's Radiophysics Laboratory. 'Applicants should possess a University degree in science or engineering,' it read. 'A knowledge of, or experience in, high frequency oscillations would be an additional qualification.' The words leaped out at me. Thanks to the direction in which Bailey had led me, the advertisement fitted my case exactly. I consulted Bailey about it. To my surprise his face immediately darkened, and he looked quite angry.

'I do not recommend you to apply for a job there,' he said emphatically.

'Why not?' I asked, taken aback.

'The man in charge of the Laboratory is Dr Martyn, and I would not like to see my best student working under him.'

I was shocked that Bailey's enmity with Martyn should extend so far. But he insisted that the Laboratory could not run properly under the control of such a man. He also expressed surprise that I was not satisfied with the work I was doing in the Physics Department. The situation was becoming awkward: I did not want to go against the wishes of the Prof, to whom I felt I owed some loyalty.

However, the matter did not rest there. Bailey wrote to someone at high level in the CSIR Headquarters in Melbourne, and apparently, as he hinted darkly to me, much correspondence ensued. I squirmed with embarrassment when I thought of the fuss being stirred up because of one small female who might or might not apply to join the Radiophysics Laboratory. However, it turned out later that there were more important issues involved.

After a while Bailey began to look more cheerful and said that progress was being made. Finally, one day, he came to me with a gleam of triumph in his eyes and announced that he had received assurances that Dr Martyn would soon be leaving his post, and that I could apply for an appointment at the Radiophysics Laboratory with his (Bailey's) blessing and support. Nonplussed, I submitted my application, and was soon accepted. My M Sc thesis would have to be completed in my spare time, but I felt optimistic that this could be achieved. In fact it took about two years.

In June 1941 I presented myself expectantly at the police-guarded entrance to the Radiophysics Lab (usually referred to as RP). One of the guards, who were called Peace Officers, conducted me to the General Office, a large, well-lit, cheerful-looking room. Four or five girls were sitting at desks equipped with typewriters and the other paraphernalia of a central office. When I gave my name they all stared at me with expressions of surprise. One of them, Sylvia Mossom, told me some time later, when we had got to know each other well, that there had been so much talk and correspondence about this scientist called Joan Freeman—who might be coming, and then she wasn't coming, and then she was coming, and so on—that they had built up in their minds the image of a large, very important-looking person. 'Imagine our surprise,' she said, 'when you finally appeared and turned out to be a slight, somewhat timid, and very gentle little person, looking quite incapable of stirring up any sort of fuss.' She giggled in her attractive, characteristic style. 'And all the better for that,' she added generously.

They were a very fine, intelligent and capable group of women, much the same age as myself, I judged. Even Gwen Miller, the senior member, who was secretary to the Chief of the Division, looked to be barely thirty. I soon found that they formed the hub of the day-to-day business of the Laboratory. There was an Administrative Section, shared with, and located in, the National Standards Laboratory (NSL), which occupied the other half of the building. But since the latter was not a classified area, a locked door prevented NSL staff from entering the Radiophysics Lab. Thus the RP Office enjoyed a considerable degree of autonomy, and discharged most of the ordinary administrative as well as secretarial functions. With this regime the Lab operated very efficiently, I discovered.

I got to know the Office staff well. With the traditional segregation of the sexes still firmly entrenched, it was natural for me to join them at lunchtime. Their companionship was very valuable to me, and no doubt I also benefited from having my ear close to the ground as far as general RP activities and gossip were concerned. Four of them—Gwen, Sylvia, Sally Atkinson and Joan Humphreys—became my lifelong friends. Sally stayed with Radiophysics until her retirement. With her attractive, vivacious personality, sound judgement, and intense loyalty, she made significant contributions to the Lab in her capacity as personal secretary to successive Chiefs. After retirement she put her extensive knowledge of RP's staff and activities over the years to good use by taking on the job of Honorary Archivist, and she is an invaluable provider of information and help to the many people, including the present author, interested in recording some of the Radiophysics Laboratory's outstanding history.

Returning to my first day at RP: I entered the Chief's office, to pay my respects and receive my orders, wondering what kind of reception Dr Martyn would give me. But he greeted

me warmly, though he was vague about what I was supposed
to do. Emphasising the necessity for maintaining the utmost
secrecy, he arranged for someone to initiate me into the
details of the Laboratory's function—the development of radar, or RDF (radio direction finding) as it was then called.

The man who acted as my guide was on temporary second-
ment from the Army as a liaison officer. He certainly did not
seem to be over-busy. He showed me all round the Lab,
talking at length; he drove me to various distant outstations
around Sydney, returning via the odd cocktail lounge; and he
even tried to organize a flight for me as an observer in an
RAAF bomber which was being used to test some airborne
radar equipment. However, this fell through, the reason given
being that a parachute was not available in my size. What an
extraordinary introduction, I thought suspiciously. I wasn't
surprised when he left the Lab soon afterwards. However,
piecing together what he told me about radar, and what I
subsequently learned, revealed a fascinating story.

The technique was first established in Britain, in 1935, by
R A Watson Watt. The principle had evolved from the method
used to study the ionosphere by reflecting pulses of radiowave
radiation from it. The remarkable fact which Watson Watt
and his small team demonstrated was that an object as tiny as
an aircraft could produce a detectable reflection. A fast time
base on a cathode ray tube was initiated by the very short
duration transmitter pulse, and a small blip, or echo, as it was
called, was recorded when the reflected pulse returned. The
time interval gave the range of the aircraft, and the direction
of the transmitting aerial gave the bearing. The radar systems which were developed from this discovery, in close
secrecy, proved to be one of the most decisive factors in the
ultimate outcome of the war.

By 1939 a chain of early-warning radar stations had been
set up along the eastern and southern coasts of Britain, and
airborne radar sets were also operating successfully. Early in
that year it was decided that Commonwealth countries should
be informed of these developments and the Australian Government was invited to send a scientist to Britain to learn the
secrets of radar, and then to set up its own research centre in
Australia. D F Martyn was chosen for this secret mission. On
his return in August 1939, the CSIR Division of Radiophysics
was born, with Martyn as Chief, and the Radiophysics Advisory Board was set up with Professor Madsen as Chairman.
Madsen, already Chairman of the Radio Research Board,

Figure 5.1 Some of the delegates to a conference at Sydney University (1949). In the front row of four: second from the left is Professor Sir John Madsen; fourth from left, Dr D F Martyn. In the middle group of ten, second on the left is Dr E G Bowen, Chief of the CSIRO Division of Radiophysics from 1946; third, Professor V A Bailey; fifth, F J Kerr; and on the extreme right, W L Price, who had been Demonstrator in Physics at the Sydney Technical College while the author was a student there in 1933/34. (Courtesy of CSIRO Archives)

played a vital part in the early establishment of the RP Laboratory, which was located in his Department of Electrical Engineering until its own building became available, and which drew its initial scientific staff mainly from the Radio Research Board. RP moved into its own quarters in March 1940. Initially working on the sample radar equipment which Martyn had acquired from Britain, the Lab was soon developing its own versions, adapted for Australian defence requirements.

My attentive guide showed me various prototype $1\frac{1}{2}$ metre wavelength radar systems which had been made for the detection of ships for coastal defence (SH D), for gun-laying (GL) to replace optical range-finders, and for airborne detection of surface vessels (ASV). The achievements of just a few young men in such a short time seemed remarkable, and I felt overawed by the sophisticated electronic developments. I might be a so-called "expert" on the properties of magnetrons and klystrons, but was still woefully ignorant about more standard types of radio equipment. My experience of radio valves had not extended far beyond an ancient-looking triode that I had met briefly in our third-year practical class.

By the time I joined the RP staff there were about thirty Research Officers. The longer-standing members seemed intensely engrossed in their various activities and I was too shy to approach them at first, though a strong sense of camaraderie was to develop before long. One of them in particular, H C (Harry) Minnett, a very competent engineer, became a good friend and colleague. He ultimately rose to the position of Chief of the RP Division, in the post-war era. Dr Pawsey, my rescuer at my klystron talk, was away on an overseas mission. He had, I gathered, played a prominent part in the development of the radar systems I had seen, as had another senior staff member, Dr J H (Jack) Piddington, an uncommunicative but impressively capable-looking man. Brian Cooper was working concentratedly with him.

I was initially taken under the wing of F J (Frank) Kerr, an M Sc graduate from Melbourne University. He was a tall, good-looking, very knowledgeable young man, with a gentle and kindly disposition. He went out of his way to be helpful to me and to give me the benefit of his experience. I especially appreciated the occasions on which Frank took me with him on operational tests at some of the Lab's field stations, particularly that at Dover Heights—a cliff-top Army site just a mile along the coast from where I had lived in Vaucluse. It was an area of rough terrain overlooking the sea, nostalgically reminiscent of the region I had enjoyed exploring as a child. The only building was a primitive concrete blockhouse on which the radar aerial was mounted. There I learned to judge the tonnage of the ships passing to and fro and to compare the size of the echoes observed on the radar screen. Frank taught me a great deal in that early period and did much to restore my self-confidence.

I was not the only woman among the research scientists.

Figure 5.2 The historic Dover Heights site where early radar tests were made and where, at the end of the war, Australian radioastronomy began. (Courtesy of CSIRO Archives)

Soon after my arrival came Ruby Payne-Scott, making a strong impact with her distinctive, very positive personality. I hadn't been aware of her existence before, but discovered that she had graduated from Sydney University with first class honours in maths and physics six years before I did. She gained an M Sc in Physics and a Diploma of Education, spent two years in the Cancer Research Department, and then one year school teaching, before taking a job in 1939 as a radio engineer in the firm AWA. She was tall, solidly built, with straight fair hair, a strong-minded, no nonsense disposition, and a shrill voice which she could use very effectively in an argument. But at the same time she had a sincere, kindly, and generous nature, to which I instinctively warmed. She was a vociferous member of the pro-communist group, which was fashionable at that time, and tried to influence me in that direction, though without success: I was not at all politically minded. On Saturday mornings she often appeared with a huge back-pack, and would stride off purposefully after work for a weekend camping expedition in the mountains. On the following Monday, she would return with a lurid tale of some hazardous experience she had had on the trip. She certainly added colour to RP, as well as making substantial contributions to its work.

Ruby also had strong feminist views, much more so than I. Yet on one occasion, when we both met a case of obvious sex discrimination, it was I rather than she who got angry about it. The episode concerned the sending to RP of many secret radar reports from Britain, on microfilm. Initially the Radiophysics Lab had no photographic apparatus suitable for reproducing these films in readable form, so arrangements were made for the photocopying to be done in the Public Library in the centre of the City. Because of the highly secret nature of the material, it was necessary for a responsible member of the RP staff to take the microfilms personally, to sit in the darkroom while the operator did the enlarging and copying, and to ensure that everything, including any reject sheets, was collected up and brought safely back to RP. This was

Figure 5.3 Ruby Payne-Scott, Alec Little, and "Chris" Christiansen at one of the Radiophysics Field Stations (c 1949). (Courtesy of W N Christiansen)

obviously a tedious and time-consuming job, which was going to stretch over a number of weeks. Ruby and I were asked to alternate in carrying out the task. I hated it. I had to sit for two hours per session in a room illuminated only by a faint red safety light, with nothing to do but watch the girl who was carrying out the job. She was pleasant enough, but I got heartily sick of two hours of her chatter, and didn't see why some of the male members of the staff in the same junior grade as myself shouldn't take a turn. But Ruby wouldn't support me in making a fuss. She said she didn't mind the exercise, and was able to knit in the dark. I had never learned to knit, and did not intend to start now. However, this was the only occasion at RP on which I experienced obvious discrimination on account of my sex. I was in general accepted on equal terms with the men.

Towards the end of 1941 several significant events took place. One of these was Dr Martyn's removal from the scene, as predicted by Bailey; he was seconded to the Army Operational Research Group. No one seemed particularly to regret his going. Although his scientific ability was unquestioned, he was said to have been indiscreet, quarrelsome, and a bad manager. The position of Chief of the Division was taken over by Dr F W G White, a distinguished scientist who had already been seconded from New Zealand to serve as Deputy Chairman of the Radiophysics Advisory Board. He had a very different personality: strong and balanced, with great drive and determination. He commanded the respect and cooperation of the whole staff and became a powerful and effective leader. I held him at first in some awe, but learned to appreciate his very fair and realistic approach, and his occasional words of encouragement.

Another event in that period, particularly important to me, was the appearance of Dr J L Pawsey (whom everyone affectionately called Joe), back from his overseas mission in America. Probably thinking of our first encounter, he greeted me with a warm, ingenuous smile. I sensed a quickening spirit throughout the Lab. Although he had a quiet, gentle presence, Pawsey's personality and influence seemed to reach out to everyone; his natural enthusiasm and drive were unbounded and infectious. I soon fell under his spell and found myself learning steadily from his example and thriving on his encouragement.

Pawsey's main errand in America had been to visit their

radar research centre, the Radiation Laboratory at MIT (Massachusetts Institute of Technology), in order to learn about the new microwave radar operating at a wavelength of ten centimetres. This important development had begun in Britain in 1940 and was disclosed soon afterwards to the Americans. Access to this very short wavelength had been made possible by the invention, early in 1940, of a new type of magnetron by H A Boot and J T Randall, working at Birmingham University under the Australian physicist, Professor (later Sir) Mark Oliphant. It was called the resonant cavity magnetron—a dramatic advance on the old split-anode type which I had used with Bailey. It could operate on a much shorter wavelength and transmit a large amount of power. It was one of the most important developments in radar in the war. For a number of applications 10 cm radar had great advantages over the longer wavelength systems.

The receiver of the 10 cm system used as its local oscillator a reflex klystron, an ingenious British version of the original device which I had read up and lectured about. This, together with a crystal rectifier, constituted the first stage of the microwave receiver.

Pawsey's major project on his return to RP was the development of a microwave radar for Australian use. A small group of staff was selected to work on the various components of the system: the magnetron transmitter; the modulator which provided the brief high-voltage pulses for it; the aerial system; the klystron local oscillator and crystal mixer; the intermediate frequency amplifier; and the display system. At first, imported magnetrons were used, but later the Lab's small but effective valve-manufacturing group succeeded in the technically difficult task of making them. Because, thanks to Bailey, I had some previous knowledge of klystrons (albeit only theoretical), Pawsey gave me the job of working on the klystron local oscillator. It was the beginning of an exciting period for me. I gained tremendous satisfaction from the sense of being a significant member of this small but important group. Pawsey gave us a superb introductory course of lectures: on aerials, electromagnetic wave transmission, and the properties of wave guides and cavity resonators, providing us in clear and simple terms with all the background knowledge we needed to pursue our development work. I was distinctly nervous of my project at first, but Pawsey was very helpful, stimulating me in his inimitable style to think for myself, and encouraging me to build up my self-confidence.

Figure 5.4 Dr Joseph L Pawsey. (Courtesy of CSIRO Archives)

One day Pawsey was observing me as I worked with a low-voltage power supply. Suddenly he said, gently, 'You are much too timid about handling voltages. Let me give you a demonstration.' He found a Variac, which is a continuously variable mains voltage transformer. There is a large dial at the front, graduated in volts from 0 to 240, with a wheel and pointer which can be rotated to give any output voltage in this range. Taking the two output leads, which ended in bare wire, Pawsey asked me to hold one in each hand while he turned the voltage up slowly, starting from zero. 'You can let go as soon as you start to feel anything,' he said cheerfully. I did as I was bid, having complete faith in him. When I started to feel a slight tingling sensation in the hands, I let go: the reading was about 90 volts. 'There you are,' he said 'There's no need to be scared of voltages lower than that.' He grinned broadly, looking rather like a mischievous schoolboy who had just carried off a successful prank. But I realised that there was wisdom in his action. He had given me a practical if unorthodox lesson in overcoming my complex about electrical voltages.

Meantime the war situation had taken a dramatic turn, underlining the urgency of our project. The Japanese, having bombed Pearl Harbor on 7 December 1941, were driving down through South East Asia at a horrific rate. By February 1942 Singapore had fallen, and the enemy seemed poised to invade Australia. Air-raid shelters were prepared and a blackout was imposed, with black paper stuck over windows, and metal slats shielding the headlamps of trams and the few cars still on the roads.

The blackout was a great nuisance to me since my eyes had an unusual defect, being very slow to become dark-adapted. I well remember one evening when a few of us were working late with Pawsey on a problem which looked as if it might take some time to solve. He suggested that we take a break, and have a meal at the Students' Union. It was still light as we walked briskly over to the Union Building and mounted its long flight of steps. When we emerged again it was quite dark. The men took the steps in their stride and disappeared into the blackness. I couldn't see a thing, and had to feel my way down step by step, watching out for the level section in the middle, and then groping by slow degrees back to the lab, seething with frustration. The others were already hard at work.

Pawsey pressed on tenaciously until the problem was solved; then we set out for the nearest tram stop, on City Road. A sudden exclamation from Pawsey announced that the exit gate was unexpectedly shut and locked. The moon had risen, and I could see the gate, much taller than I was and topped with closely spaced, vicious-looking iron spikes. Pawsey with his long legs had soon managed to climb over. The others were also able to negotiate the obstacle, using a horizontal bar halfway up as a toehold. They went straight on to the tram stop. I tried the same method, but my legs were not long enough to reach over the top. What should I do? The Parramatta Road exit was a long way away. So I tackled the gate using a higher bar, close to the spikes. Balanced precariously, I had just about completed my manoeuvre when it occurred to Pawsey to come back and help me. I promptly lost my balance and sat painfully on the spikes. Acutely aware of the oncoming bruises, as we waited for the tram, I reflected that, as usual, I had been treated as "just one of the boys": Pawsey's reaction had been very much an afterthought. But I appreciated this lack of distinction on account of my sex.

During late 1941 and early 1942, a succession of events occurred which highlighted the importance of the RP Lab for Australian defence. Until Japan entered the war, RP had concentrated on developing Shore Defence (SH D) and gun-laying (GL) radar against possible sea attack, but after Pearl Harbor the necessity for a long-range air-warning system became urgent. In December 1941 a group led by Jack Piddington and Brian Cooper modified a $1\frac{1}{2}$ metre wavelength SH D set with remarkable speed to provide the maximum possible range. After successful trials, three operational sets were quickly built for the RAAF, and one of these was sent to Darwin, Northern Australia, accompanied by RAAF technicians. The latter, however, had problems with the new equipment and the set was still not working when, on 19 February 1942, the Japanese staged a devastating bombing raid on the town. After some procrastination a call for help was sent to RP. Jack and Brian set out at once for Darwin, travelling by a tedious series of hops via Adelaide and Alice Springs.

The radar set was still out of commission when they arrived, and they were tantalized by a minor air raid while they were working on it. However, by the next morning the set was operational. Staring at the radar screen, the two men

Figure 5.5 Brian Cooper (on the right) with colleagues on the roof of the Radiophysics building. (Courtesy of B F C Cooper)

suddenly noticed a blip growing out of the noise at a range of 86 miles. 'Gee! This is a raid,' exclaimed Brian. The RAAF crew alerted a squadron of American fighters based nearby, and these were able to intercept and disperse a large incoming flight of Japanese bombers while they were still about thirty miles from Darwin. Jack and Brian, watching the screen, could see it all happening. After that there was a rapid curtailment of raids on the town, and the RAAF treated radar with a new respect. Everyone at RP shared in the sense of triumph which Jack and Brian experienced.

From the Darwin set one of the most noteworthy products of RP was evolved. This was the Light Weight Air Warning set (LW/AW). It became particularly important when, in mid 1942, the Japanese advance in the Pacific was halted, and the long, bitter process of driving them back began. The LW/AW was acknowledged to be the most reliable set of its kind; it was readily transportable, and stood up to the appalling conditions under which the island-hopping warfare had to be conducted, while its heavier American counterpart tended to sink in the mud.

Meantime our microwave group was pressing ahead as fast as possible with the components for a 10 cm radar set. My klystron oscillator was performing satisfactorily, but it became evident that great stability was necessary in its high-voltage supply. A special variable, voltage-stabilized power supply was required. Pawsey said that I should design and build this. 'But I know practically nothing about circuitry of this sort,' I protested. 'This will give you a good opportunity to learn,' he replied with a smile.

I think Pawsey could have done the whole job himself in the time he gave to guiding me, but he carried on with characteristic patience, feeding me with suggestions at the appropriate moments, and then leaving me to develop them. He insisted on my doing the job logically and thoroughly, working out all the necessary theory. I did indeed learn a great deal from that exercise, and gained much satisfaction when the completed power supply worked exactly as it should. Then, at Pawsey's request, I wrote a full report on it. This was typical of the way in which Joe Pawsey operated. His own publication list is not very long; but there must be many papers, written by people who have had the benefit of his ideas, bearing the Pawsey stamp.

Other components of the 10 cm microwave set which were my responsibility were the transmit/receive (TR) switch and the crystal mixer. The TR switch is an ingenious device which permits the same aerial to be used for both transmitting and receiving. The technique was first developed in 1940, for $1\frac{1}{2}$ metre radar, by Pawsey, with Harry Minnett (and independently in the UK); it soon became universally adopted. In our case the switch consisted basically of a cavity resonator, like that of the klystron, with a glass envelope containing gas at low pressure at its centre. The powerful transmitter pulse caused a glow discharge which effectively short-circuited the cavity and prevented damage to the delicate crystal mixer of the receiver. Between the transmitter pulses the resonator could pass echo pulses, without loss, into the receiver. Developing and testing the TR switch system was the most interesting part of my microwave work.

I also made the crystal mixer, a tiny encapsulated version of the tungsten "cat's whisker" and silicon crystal of the earliest radio receivers. I had to make and test a number of these, earning myself the title "Crystal Crackin' Momma". It was fiddling work but I remember it because it was during that period that Professor Mark Oliphant visited the Lab and took a close interest in the microwave group. He was an impressive looking man, tall, of big build, with a florid complexion and a shock of prematurely white hair. He had an authoritative manner and looked capable of great drive and determination. I knew that he had gone from Adelaide to play a prominent part in the Cavendish Laboratory with Rutherford, and that it was under his direction, as Professor of Physics at Birmingham, that Boot and Randall were led in 1940 to invent the resonant cavity magnetron—the vital component of our 10 cm system. He was encouraging to me personally and I was thrilled to make contact with someone of Oliphant's eminence, from Britatin. Most Australians felt isolated in those days, and thought of Britain idealistically as the Mother Country and the hub of the civilised world. I treated with some awe the few scientists I knew who had gained Ph D degrees by going to England for two or three years. (Australian Universities did not then offer this higher degree.) But Oliphant, having become a famous professor in Britain, was worthy of my special respect.

In due course all the components of our 10 cm radar set had been made and separately tested. The moment had arrived for assembly of the whole system and the testing of its radar capabilities. This was to be carried out at the Lab's field station at Dover Heights, which Frank Kerr had introduced

Figure 5.6 One of the Radiophysics labs. Seated in the foreground is Frank Kerr in discussion with Arthur Higgs; back left is the author. (Courtesy of CSIRO Archives)

me to. It was really not necessary for everyone in the group to be at hand for the operation, but Pawsey, with typical thoughtfulness, said that we should all come and share in the fun of seeing this climax to our labour. So all our bits and pieces of gear were piled into a big trailer, and we set out for the site in a state of suppressed excitement. I stood around for a while, watching the frantic activity, but had my turn in due course, inserting the components that I was responsible for. Finally, all was ready for switching on. The adjustments went smoothly, and soon we were looking out for passing ships to act as targets, peering over each others' shoulders for a glimpse of the radar screen, and shouting with excitement when echoes started to appear. The set proved a great success, picking up a 6000-ton ship at a range of about 45 miles, which exceeded expectations. Pawsey was grinning with boyish delight, and I felt a great sense of exhilaration.

The Australian Navy was duly impressed with the set's performance; many of their ships were ultimately equipped with the microwave radar, which gave valuable service. It was also used effectively by the Army for coastal defence.

No sooner had the prototype 10 cm set passed into the production stage than our microwave group was launched into an even more ambitious project. This was the development of an entirely new radar system, operating at a wavelength of 25 cm. Pawsey perceived that this was the optimum wavelength for air-warning and height-measuring equipment, effective against both high-flying and low-flying aircraft. It was a big project for our lab to undertake, since such a system had not been developed either in Britain or America. Many unique components had to be designed for the new wavelength. The magnetron and klystron, particularly key items, were taken on by the Valve Lab under the Melbourne Professor, L H (later Sir Leslie) Martin, with impressive success. The Transmit/Receive switch was assigned to me and Harry Minnett. This was an exciting challenge. I had by then advanced considerably in self-confidence and research ability, and was able to inject some ideas of my own into the design.

Progress all round on the 25 cm radar was slow, with so many new techniques to be developed and problems to be solved, but finally a prototype which met all specifications and expectations was achieved. It was a triumph for a small laboratory like ours, but ironically, with the unexpectedly sudden end to the war in August 1945, the system never saw active service.

Although throughout the war period everybody in the Lab concentrated intensely on their work, with a sense of common purpose and immediacy, life there also had its lighter moments and diversions. We were a young group, mostly in our early to mid twenties, almost all single, and with exuberant spirits which needed regular outlets. Our every day activities were frequently punctuated with friendly banter and spontaneous bouts of fun, as well as planned amusements. The pre-Christmas party in the Lab became a regular feature, and, particularly during the latter part of the war when the tension began to ease, external functions and excursions were organized by willing volunteers. I remember well a hilarious fancy-dress dance held in the Students' Union—a challenge to the ingenuity of the participants in the face of war-time austerity; a sports meeting on the playing field opposite the Physics Building, with fiendishly devised team competitions; and a very successful trip to Avalon Beach, achieved somehow on hoarded petrol rations, with a sand-castle building competi-

Figure 5.7 Radiophysics Lab excursion to Avalon Beach. Standing, left to right: Ron Bracewell, Gwen Miller, Sally Atkinson, Tina Rebane. Sitting: Joan Humphreys.

tion, the prize for which was a barrel of beer (evidence of the strong male influence). Being a member of the winning team, I was expected to take my share of the prize. I survived the experience, though it put me off beer for some time.

The veil of secrecy tended to keep us rather to ourselves, and many of our social activities were contained within the RP group. However, for some occasions we combined with the Standards Lab, where quite a number of women were employed. In the Metrology Section in particular, a large group of girls was engaged in high-precision measurements for munition components, and they provided an attractive means of balancing the sexes.

Apart from the organized social events, there were plenty of unrehearsed incidents in the Lab to provide general entertainment. I remember one episode, which must have occurred quite early on, involving Ruby Payne-Scott, myself, and the lady in charge of the Library. This, like the Administration Section, was shared with the National Standards Lab and located in the NSL building. The Librarian was a strong-willed, none too popular lady of advancing years, with a strictly conventional outlook. She took exception to the fact that Ruby sometimes wore shorts in the lab, and appeared thus attired in the Library from time to time. Many of the men also wore shorts, but that, apparently, as far as the

Librarian was concerned, was different. She wrote a peremptory note to Ruby, requesting that she not wear shorts in the lab. Nor was I exempt from her disapproval. I was requested not to smoke in the lab—a habit I had acquired during my later University period.

Ruby was very angry, declaring that shorts were more respectable than a skirt while she was climbing on a stool or working on the RP roof or in the field. She continued to wear her shorts, and urged me to carry on smoking, since most of the men did: a matter of principle was at stake. All the RP staff barracked for us enthusiastically.

In due course we both received from the Librarian a summons to attend a meeting in the Library, at which the general question of the proper attire and behaviour of women in the lab was to be discussed. Ruby deliberately changed into shorts for the meeting, while I decided to boycott it. Ruby was soon back. She had been promptly ejected. The battle of wills continued, being resolved only when the Librarian retired, and was succeeded by Marjorie Barnard, the well-known Australian author. She was an interesting and very enlightened woman, who was not in the least concerned about Ruby's shorts or my cigarettes.

There was one case involving sex discrimination, beyond the powers of RP to control, which Ruby went to extraordinary lengths to circumvent. It was in 1944 that she let it be known that she was living with a man (Bill Hall) to whom she was not married. Nowadays little would be thought of such a situation, but in the 1940s "living in sin", as it was called, was looked on askance. However Ruby, who had always kept her private life very much to herself, carried on as usual, unperturbed. It was not until some two years later that the truth came out: she had been married to Bill all along, but did not want the fact to be known officially because of the long-standing rule in Government Establishments that married women could not be employed on a permanent basis; they could be given only temporary appointments, renewable annually. Ruby had hoped, by her deception, to evade what she considered to be an outrageous and discriminatory law. All her RP friends, having developed a strong affection for Ruby as well as respect for her scientific abilities, greeted the story with hilarity, and sympathized with her attitude.

During 1943 there was a significant influx of new recruits to RP, including recent graduates from Electrical Engineering, some of whom proved to be a particularly jolly lot. One of these was R N (Ron) Bracewell, small and slight, with a strong sense of humour, and a capacity for leg-pulling. He was extremely bright, and had an encyclopaedic range of knowledge: he could discourse eloquently (though sometimes with tongue in cheek) on subjects as diverse as metaphysics, ancient history, and cryptography. Apart from his entertainment value, Ron was also a very able scientist, with original ideas and particularly strong theoretical ability.

Another of the new recruits was R B (Bob) Coulson, a tall, personable young man with a strong, outgoing personality, a ready wit, and an air of levity which belied his not inconsiderable engineering capabilities. He and kindred spirits, like the loquacious mathematician from Melbourne, John Ryan, who did his serious work at night and tended to relax in the daytime, had a distinct effect on the level of frivolity in the Lab. Pawsey used to get slightly impatient when high spirits

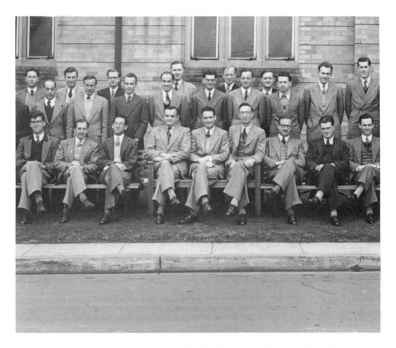

Figure 5.8 Some of the Radiophysics Lab research staff (1952). Front row, left to right: B F C Cooper, H C Minnett, J Warner, L L McCready, E G Bowen, J L Pawsey, J H Piddington, P Squires, R N Bracewell. (Courtesy of CSIRO Archives)

got a bit out of hand, and was heard to mutter something about people who were a disruptive influence. I also remember Len Hibbard, another of the more senior staff members, remarking half in jest, half in earnest, that there were two ways of pronouncing the word "laboratory". In American usage the accent was on the first syllable; in English, on the second. 'But it is clear,' he continued, 'that the accent should be on the "labor", and not on the "oratory". However, I doubt that our work suffered seriously from these diversions. As the war outlook improved, it was inevitable that we should relax a little.

There was, however, one event which probably did disrupt our work somewhat. This was a grand Musical Revue, staged jointly by Radiophysics and the National Standards Lab. Master-minded and directed by Bob Coulson, it was called "Hush Hush", and brought to light a remarkable range of talents in music, drama, dances, and humorous sketches. It included a variety of skits, including one devised by Ron Bracewell, using the music of *The Mikado*; this featured a "Lord High Clerical Officer", liberally draped with red tape (the head of the Administration Section, based in NSL, being regarded as fair game for such a parody). I enjoyed taking part in a ballet, "Blues in the Night", with Sally, Sylvia, and other members of the RP Office; also a slick song and dance sequence, to the music of *Cornsilk*, representing eight land-army girls. The other seven girls were largely from NSL's Metrology Section. One of these was Pat Eade, a pretty, fair-haired, vivacious girl, whom later I got to know well, as Mrs Brian Cooper.

The Revue was altogether a great success and a fillip to the Lab. With it, and a second one which he staged a year later, Bob Coulson established beyond doubt his organizing abilities and his flair for handling people. It was not surprising to me that, later in his career, he became Managing Director of the English Electric Valve Company, in Chelmsford, where he and his family became close, long-standing friends of mine.

In connection with the Revue, the story is frequently told at RP about a young girl who had just been appointed to a junior position in the RP Office: she was auditioned for a singing part, but was rejected. She left RP before long to concentrate on her music studies, and later became a world-famous opera singer. Her name was Joan Sutherland.

Early in 1944 an important and colourful personality entered the RP scene. This was Dr E G Bowen, well known for having been one of the first scientists to join Dr Robert Watson Watt, the inventor of the radar technique, in the historic work, beginning in 1935, which was to have such a decisive influence on the course of World War II. Bowen, popularly known as Taffy, had some graphic tales to tell us of the early struggles of their tiny team in getting their systems operational and accepted by the Services. Then, in 1940, he was the man who, following the British decision to divulge the radar secrets to America, was entrusted with the first cavity resonator magnetron to cross the Atlantic. As part of the secret mission led by Sir Henry Tizard (scientific adviser to the Ministry of Defence), he contributed to the establishment of the Radiation Laboratory and worked with the Americans there, as well as the British Liaison team. Then, at the end of 1943, Bowen was persuaded by Tizard to move on to Australia. At RP he started off as Deputy to Dr White. The latter soon left to join the CSIR Executive, ultimately becoming Chairman, and Taffy was made Chief of our Division, a position he was to hold very effectively until 1971. His lively figure and easy manner soon became familiar to everyone as he moved enthusiastically around the Lab, getting to know in detail what was going on, and giving us the benefit of his broad experience. The timing of his arrival was opportune: he was able to exert an important influence as the activities of RP began to enter a new phase.

By the end of 1944 the war situation had changed dramatically. The Allies had gained the ascendancy as the Japanese were pushed back, and much of their shipping destroyed. Although a costly struggle, including a possible invasion of the Japanese mainland, was still envisaged, the successful conclusion of the war was in sight, and the radar equipment already in use, or at the development stage, was regarded as adequate for remaining wartime needs. So, although the Lab continued to have commitments to the Armed Services, its function of devising and developing new radar systems was no longer required. Bowen therefore began to address the question of the Lab's future work.

Calling us all together, he made the point that RP, with its concentration of talent (the professional staff by then were about sixty strong), broad experience, and first-class equipment, was in an excellent position to move into fields of peace-time research, both fundamental and applied, which looked promising. Planning should begin at once, he said. Already some applications of radar to civilian life were being considered: Pawsey, with a sizeable group, was studying the effects of the atmosphere on ultra-high frequency radiowave propagation; and Bowen was enthusiastic about establishing a group to study cloud and rain physics by radar techniques. He urged us all to think about producing ideas for other possible avenues of research which he could present as part of an overall proposal to the CSIR Chairman.

Thus stimulated, I began to ponder on the work I was currently doing with the TR valves for our 25 cm radar system. The immediate problem was the choice of gas filling to give the best operational performance—a rapid, low-field electrical discharge due to the magnetron transmitter pulse, and a fast recovery for the subsequent reception of the radar echo. All my tests had to be on a trial-and-error basis, because very little in the way of experimental data, and no reliable theory, existed concerning the behaviour of low-pressure gas discharges at microwave frequencies. I could see that there was a need for a systematic study of the subject. So I decided to write a brief note to Bowen, pointing out the desirability and scientific interest of such an investigation. Bowen seemed delighted with my initiative, and suggested that I give a talk to the Lab about my proposal.

Though I did not realise how important it was going to be for me, I prepared my talk carefully. Drawing on the knowledge I had gained in my honours year from Bailey's course on electricity in gases, and the work I had done for my M Sc thesis, I began with a simple introduction on the basic atomic processes involved in the production of a glow discharge by a high-frequency electric field; I pointed out the differences in behaviour expected with the very high microwave frequencies; I described the small amount of work that had so far been done; and I outlined the proposed experiments to elucidate the processes, using 10 and 25 cm wavelength fields. I concluded with a vague reference to possible applications.

I felt that it was a good talk. I had quite a large audience, and I obviously took them by surprise. They were all impressed, even Pawsey. I suppose it was because none of them knew much about the topic, and found my talk unexpectedly interesting. Bowen said, very enthusiastically, that I should go ahead with the proposed experiments, and would be provided with adequate technical support. He incorporated my proposal in the memorandum he submitted to the CSIR Chairman about RP's research programmes in the post-war era; this was readily accepted.

The end of the war came, of course, much sooner than expected. When the news broke about the atomic bombs dropped on Hiroshima and Nagasaki, almost all of us were completely mystified. John Gooden, an outstandingly bright young physicist from Adelaide University, who had been at RP for a year or two, had some inkling about nuclear fission, but it was some time before I understood what it was about. At the time, there was no thought of the moral issues connected with the dropping of the bombs, but simply an enormous sense of relief that the event had brought the war to a quick end, and thereby saved the lives of many thousands of Allied Servicemen. It was a strange sensation, though, to find our wartime activities so abruptly terminated. Bowen, with his astute managerial instincts, decided that the Lab should busy itself by preparing to give, as quickly as possible, a series of public lectures divulging the details of the secret work on which we had been engaged. Many of us were involved in this—I was responsible for the talk on local oscillators, particularly klystrons. The lectures were a great success, so Bowen decided that our papers should be published in a book, which he edited, called *A Textbook of Radar.* This, the first such publication anywhere in the world, appeared in 1947.

Meantime, the Lab was reorganized for its peacetime activities. Extra accommodation was needed for some of the new projects, including my gas-discharge work, and two large rooms in the University's Electrical Engineering Department were allocated to us for this purpose. This section, called the Vacuum Physics Laboratory, was placed under the supervision of Dr Owen Pulley, a member of RP's senior staff, who was a capable engineer and a likeable man, though not with Pawsey's dynamism. My experiment was going to require a considerable amount of equipment: a magnetron oscillator plus power supply, a vacuum system, gas-handling apparatus, gas-discharge chamber, and various measuring instruments.

At this stage I suffered an acute attack of cold feet. It was the first time that I had initiated a new project, and this was one that I would have to carry out on my own. I wouldn't have Pawsey standing by to encourage and advise me. No one at RP was in a position to help me except with the technical assembly of the equipment. I could talk to Bailey about the physics of the problem, but he was basically a theoretician, and would not be able to advise me on the practical aspects. I will probably make a hash of the whole thing and look ridiculous, I thought anxiously to myself. Why had I let myself in for this, instead of trying to get into Pawsey's radiowave propagation group, where I could have worked comfortably under his guidance? But I had committed myself; there was no turning back now.

Fortunately, I had the technical support of an extremely pleasant and capable lad called Alec Little. Alec's personal story is a remarkable one. In 1941 he joined RP as a junior office boy straight from school, but, because of his evident practical ability, he was soon taken on as a technician in the workshop. Before long he had reached the status of Technical Assistant in the lab, where he gave invaluable support to the scientific staff and acquired a wide range of experience in a number of radar projects. I owe him a great debt for the vital contribution he made to my work in obtaining and assembling all the necessary equipment, a job he did with characteristic efficiency and cheerfulness. He was to have an outstanding subsequent career, gaining a University degree, and ultimately becoming a much respected Assistant Professor in the University Physics Department. Thanks to Alec's help, I was able to get my experiment going, and, in spite of making a number of mistakes on the way, through lack of experience, I ultimately achieved some satifactory results at the 10 cm wavelength.

Meanwhile, other projects going on in the Vacuum Physics Lab were proving quite dramatic. They were concerned with the application of microwave techniques to the acceleration of elementary particles—electrons or protons. With a 10 cm magnetron transmitter and a resonant cavity, electron energies up to a million electron volts were achieved. The device proved also to be a veritable x-ray machine, causing considerable excitement and not a little dangerous radiation.

The chief protagonists in this pioneering work were an interesting group of people, and I enjoyed the stimulation I gained from their proximity. But I was unhappy with the geographical separation from the Radiophysics Lab, and missed the regular company of all my old associates there, particularly Joe Pawsey. I dropped in whenever I could, and learned about Pawsey's new exploits. I knew that, well before the end of the war, he had been itching to look for what was called "cosmic noise": extraterrestrial radiowave radiation, which had first been reported in 1933 by Jansky, in the USA. In 1944, although the war was still on, Pawsey made some observations with standard radar equipment on the roof of the

Radiophysics Lab building. Later, as opportunities for concentrated effort opened up, he began to measure radiowaves from the Sun. I remember vividly his greeting me one day in great excitement, exclaiming: 'The temperature of the Sun's surface is a million degrees.' Then, at the Dover Heights station, he started to observe enhanced solar noise attributable to sunspot activity. He devised an ingenious technique, using the surface of the sea as an interferometer, to show that the noise did indeed come from the sunspots. Ruby Payne-Scott was one of the people working with Pawsey on these measurements. Infected with his bubbling enthusiasm, she stirred in me some pangs of jealousy as I contemplated my own unspectacular efforts on gas discharges.

Those solar noise observations were clearly interesting, but little did I imagine how dramatically the work would develop. I doubt if Pawsey himself envisaged at that stage the huge and exciting field of research, to be called Radio Astronomy, into which these small beginnings were to evolve. Under his inspired leadership, the Radiophysics Lab, in the course of a few years, not only extended greatly the work on solar activity, but embraced noise from the Milky Way, the identification of discrete radio sources, and a host of fascinating new phenomena. Australian physics was placed squarely on the international science map. The only competitors to begin with were a parallel group at Cambridge, under Dr (later Sir Martin) Ryle.

At some stage during my period in the Vacuum Physics group, Professor Bailey offered me a permanent lectureship in the University Physics Department. I was gratified by the offer, but could not contemplate it seriously. The thriving research atmosphere at RP, the stimulation to be derived from a large number of dedicated fellow-scientists, the effective management of the Lab by Bowen, and the pervading influence of Pawsey, to whom I hoped to return when the gas-discharge work was complete, all combined to make the Lab irresistibly attractive to me. By contrast, the University Physics Department was much smaller, and the teaching responsibilities, I realised, would inevitably tend to take precedence over research activities, which I considered at the time to be more satisfying. So I turned down Bailey's offer.

But then an unexpected event occurred which was to change the whole course of my future life.

Chapter 6

Cambridge

One day, early in 1946, Bowen suddenly called me into his office.

'Look at this,' he said, with a cheerful glint in his eye.

He handed me a Memorandum which had come from the CSIR Head Office in Melbourne. It announced that the Council had decided to award a limited number of Senior Studentships to meritorious young staff members, for the purpose of training in Britain, at a University of their choice, with the opportunity of studying for a higher degree.

'These Studentships are of course open to all Divisions of the CSIR,' Bowen pointed out. 'I can't say what your chances would be, but I think that you and Ron Bracewell should apply.'

I was dumbfounded. The thought of being able to work for a Ph D degree in England—an opportunity that had in the past been afforded only to an exclusive few Sydney University students—had been for me only a pipe-dream. Yet here on offer was a sporting chance for me to be sent to England—perhaps to the revered Cavendish Laboratory in Cambridge. I was also astonished that, with all the talent available at RP, Bowen should pick me out in this way. The choice of Ron I could understand: he had shown evidence of great theoretical as well as practical ability. But why me? Apparently my initiative in starting up an original programme of research on gas discharges, whether or not dramatic results were coming out of it, had really impressed Taffy. I felt gratified, though undeserving.

But then the full impact of what such an enterprise would mean, if it came off, began to dawn on me. How could I leave my mother like this after she had devoted the best years of her life to me and was accustomed to our close companionship? I had never been away from her for more than the odd week's holiday. Could I suddenly leave her on her own for two or three years?

As with everything that happened to me, I shared this new turn of events with her that evening. Far from looking shocked, her eyes lit up with genuine excitement. 'This would be just what you have dreamt of,' she said. 'Of course you should apply.' She admitted that she would be terribly lonely without me, but said she could not contemplate ever standing in my way. She had had her young life; my success was her reward, and I must pursue my career as opportunities arose. So I applied for a Studentship, pointing out to my mother that I probably wouldn't get one anyway, since there would be candidates also from the National Standards Lab, and from other CSIR Divisions, which operated in scientific fields as diverse as chemistry, agriculture, food research, oceanography, etc.

The result came out remarkably quickly. Both Ron and I were awarded Studentships. Bowen was highly delighted: to gain two awards was for him a gratifying indication of RP's high rating in the CSIR.

I told Joe Pawsey about it at once. His face shone with pleasure. 'Where will you go?' he asked enthusiastically. 'To Cambridge, if I can,' I replied promptly. His enthusiasm was redoubled. 'Splendid! You must go to my old College, Sidney Sussex,' he exclaimed. Then, reflecting for a moment, his face fell. 'Oh, I suppose you'll have to go to a women's college,' he said in a disappointed voice.

Dear Joe, he still tends to think of me as "one of the boys", I reflected. I started to question him about the Cavendish, and he told me what a memorable experience it had been for him to spend three years there in the early 1930s, working for his Ph D. Nostalgically he described the attractions of life in Cambridge. Then he went on to talk about Rutherford, whose inspiring presence had pervaded the whole of the Cavendish, and about his supervisor, J A Ratcliffe, an outstanding experimentalist in ionospheric and radio research, and a splendid teacher.

'What subject do you want to study?' Pawsey suddenly asked me.

'Well, I've been thinking about this, and I'd like your advice,' I replied. I told him how fascinated I had become by the accounts I had heard of Rutherford's remarkable achievements in nuclear physics. To work at the Cavendish in this particular field had been something for me to dream about. But I had had little grounding in the subject, and it made more sense, in view of the experience I had gained at RP, for me to be involved in some topic related to radiowave radiation. I suggested that I should apply to work under Mr Ratcliffe, Pawsey's old supervisor; perhaps I might even be able to get involved in some aspect of cosmic noise—the exciting new field that Pawsey had initiated at RP.

Pawsey's face became serious at this suggestion, and his response was hesitant. For reasons which I never fully understood, he discouraged the idea of my going into radiowave research. In some embarrassment he pointed out that this would entail working a lot at field stations, where the conditions were rough and primitive, and there were no facilities for a woman. I immediately thought of the Dover Heights station, where I had worked occasionally, and knew what he meant. Evidently he now had my sex in mind. But he must have had other reservations about my going into such work. Perhaps he simply felt that I should be encouraged to follow my initial inclination. At any rate, he said that it seemed a good idea for me to pursue nuclear physics. I would soon acquire the knowledge I needed for a particular project, he thought. He made the point that the main purpose in working for a Ph D degree was to learn the art of research in the environment of a first class laboratory, the actual topic of the work being less important.

I was happy to accept Pawsey's advice, and Dr Bowen seemed satisfied with the proposition that I should pursue nuclear physics. He told me that Ron Bracewell wanted to work on radiowave propagation in the ionosphere, under Mr Ratcliffe, so it was a good idea for me to be doing something different.

I had next to find out if the Cavendish would accept me. I applied immediately, realising that, because of the war, there must be a six-year accumulation of students now wishing to start Ph D courses; I wondered anxiously how difficult it would be for me to gain a place. However, a favourable response soon came: the Cavendish was willing to have me provided that I was accepted by one of the Colleges. There were only two women's Colleges then: Newnham and Girton. I applied to both and they both agreed to admit me. From Rachel Makinson I learned that Newnham, her College as an undergraduate, was close to the Town Centre, whereas Girton was some two miles away, deliberately located, when it was first established, at a "safe" distance from the influences of male students. So I opted for Newnham, with great relief at having my place in Cambridge assured.

Meantime, Ron and I received notices of berth allocations in a ship called the *Orbita*, due to leave Sydney in mid August 1946. This also was a relief since normal shipping had not yet been resumed. There were just two ships available during the latter part of the year for the transport of passengers to England. They were still fitted out for troop-carrying, and I could see that it was going to be a crowded trip, though I didn't visualize how bad it was to be.

In due course some detailed instructions were sent to me from Newnham College: I should arrive at the beginning of term, the first week of October, and, because of the strict rationing still in force in Britain, I should bring with me all the personal items, clothing, bed-linen, blankets, and towels that I would need for two or three years. Just about everything except the kitchen sink, I muttered to myself, as I began to contemplate how I was going to find containers in which to pack all that plus textbooks and papers. I would even have to bring my own deck-chair for the boat, so the shipping agent informed me. My mother worked with characteristic drive helping me prepare everything for the journey. She must have felt exhausted when it was all done. Then, on 9 August 1946, came the agonizing farewell, and I was sailing down Sydney Harbour with a full heart, into a new world.

I realised from the start that the trip was going to be very uncomfortable. To begin with, a newspaper report that I had seen just before leaving said that when the boat arrived in Melbourne, after the voyage from Britain, some of the passengers were so disgruntled with the conditions on board that, during a farewell drinking session, they piled together the furniture in one of the two lounges and set fire to it. The upshot, all too evident when we went aboard, was that there was only one lounge room available for the grossly over-crowded boat. Furthermore, as a result of this incident the Captain had decreed that the boat would be completely dry for the return journey.

When the old *Orbita* was in its heyday (it dated back to the first World War at least) my cabin, with three wash-basins, had probably provided comfortable accommodation for a family. Now it contained fourteen berths in three-tiered rows around the room, still with only three wash-basins for its fourteen female occupants. The communal baths were all unusable, and the showers, only just so. There was no air-conditioning, and by the time we reached the tropics I had taken to sleeping up on deck in my deck-chair.

There was only one port of call at which we were allowed ashore after Fremantle. This was Aden—a poor, depressing place, but hailed by a number of passengers as a surreptitious source of alcohol. A group of them missed the sailing and had to hire a tender to chase the boat. Then they had to climb a flimsy rope ladder which had been lowered for them over the side. For the entertainment-starved crowd aboard it was a welcome diversion to witness the struggles of these men, bottles dropping from their pockets, as they laboriously negotiated the ladder. After that we stopped only at Port Said, for provisioning and refuelling; we had to stay on board because of the uneasy political situation there.

However, I found one great compensation for the rigours of the journey. There were on board a number of young people, like myself, whose objective was further study at British Universities and similar institutions. It was an impressive concentration of intellectual talent. A close bond of mutual interest grew up between us, and we created our own entertainment. One of our favourite occupations, especially in the evenings, was forming discussion groups to exchange ideas on every subject under the sun. Someone had the bright idea of using one of the open lifeboats, which were stowed on the top deck, as an assembly point. Ron Bracewell, I remember, played a prominent part in organizing and leading some of these meetings. Lying back, out of sight of everything except the wide starlit sky, I listened appreciatively to expositions, or joined in lively debates, on topics ranging from astronomy to religion and from philosophy to art. Thus much of the tedium was taken out of that long six-week voyage, and I look back with pleasure to those stimulating periods and to the friends I made then.

The *Orbita* docked at Liverpool, and after the long, irksome procedures for disembarking, the final blow was the requirement that passengers must collect all their hold luggage and take it away with them. My grandmother's large old-fashioned trunk, which contained all my blankets and linen, was one of the last pieces to be rolled laboriously down the gangway, to join the rest of my miscellaneous collection of baggage, including the deck-chair. The last train to London had long since departed. I was thankful that a Sydney University contemporary, a chemist called Reg Goldacre, found himself in the same situation, so that at least I had a companion with whom to consult on what to do next. First, though, I had to telephone Gwen Miller, the Radiophysics secretary who had returned earlier to her native England and was to have met me off the train in London. I located a functioning public telephone, but had great difficulty in understanding the operator, who had a strong Lancashire accent. By way of apology I told him that I had just arrived from Australia. He sounded surprised. 'But you speak almost perfect English,' he said, his dialect becoming broader than ever.

I finally got a message through, and Reg and I persuaded a helpful taxi-driver, with a commodious London style taxi, to take our luggage in relays to the railway station, where we thankfully left it in the baggage store. He then took us on a tour of Liverpool, trying to find accommodation for us. The hotels were all full: with people off a boat, the proprietors explained. As we drove around, we saw some of the bomb damage that had been inflicted on the city during the war. I was appalled at the extent of the devastation. This first-hand sight of so much rubble that had once been buildings brought home to me, more than all the newsreel reports that I had seen in Sydney cinemas, the dreadful way in which British civilians must have suffered in the war.

It was getting quite late by the time we found a seedy-looking guest house which offered us two rooms. Mine had once been a bathroom: the bed consisted of a wooden board, laid across the top of the bath, supporting an ancient mattress and bedding. An opaque panel of glass in the door seemed to suggest to other guests that this was still a bathroom, as the doorhandle was rattled at intervals during the night.

The next day I managed to get my heavy luggage sent on to Newnham College and then I boarded a train to London. To my surprise, the other occupants of the carriage soon struck up a conversation with me, and in no time had learned with interest all about me and my situation. So much for the English reserve that I had been warned about, I thought. In due course I discovered that whereas north country folk were naturally friendly and conversational, even with strangers, the reserve tended to persist in the people of southern England. Silence would invariably reign in a railway carriage in Surrey, or even in Cambridgeshire. But underneath, when I got to know them, I found that the southern English were in general very kind, helpful, considerate and sensitive, and I made many good friends among them.

After an exciting glimpse of London, and a brief holiday with Gwen in her lovely old Surrey home, I at last set out for Cambridge, on a crisp October morning—the day that Newnham College had specified for my arrival.

The approach by train gave an unexceptional first glimpse of the town. The railway station, still with its dreary wartime nonentity, differed from other English stations I had seen only in having just one very long platform, with ill-defined train-length sections called Platform 1, Platform 2, etc, much in the way that I was to find the main streets of Cambridge changing their names as they led past various colleges; for example, in the space of about 400 metres Trumpington Street became successively King's Parade, Trinity Street and St John's Street.

Emerging into the station precinct I was encouraged by the sight of a familiar, cheery face. Ron Bracewell had come to meet me and, having arrived two days earlier, was eager to act as my guide. Taking a bus to the Town Centre, we were soon walking through the busy market place and past the University Church of Great St Mary's, to emerge onto King's Parade. I was spellbound by the sudden vista. To the right stood the Senate House in its beautifully proportioned classic elegance; to the left, the richly ornate entrance to King's College; and in front of us, in all its glory, rose King's College Chapel, looking immensely tall and grand, with its delicate, tapering pinnacles pointing skywards. The interior of the Chapel, which we approached via King's spacious quadrangle, was even more impressive, with its magnificent fan vaulting, rich carving, and huge windows, though the famous stained glass, which had been removed and stored during the war, had not yet been replaced.

Reflecting on how privileged I was to be here, I walked with Ron along the Backs, absorbing the beauty of this unique Cambridge feature: the backdrop of mellow college buildings, the wide sweep of lawns, imaginatively laid out gardens, tree-lined avenues, the River Cam flowing gently under the distinctive old college bridges, and the spacious meadows beyond, extending to Queens' Road. There is still for me no more delectable place of man's contrivance, in all of Britain.

After lunch in one of the picturesque old cafes with which Cambridge seemed to abound, I made my way to Newnham College, which flanked the left-hand side of Sidgwick Avenue—named after Henry Sidgwick, founder of the College in 1871, I learned. The buildings, linked by corridors and strung out in a long line, presented a somewhat formidable façade of Victorian red brick, in marked contrast to the magnificent old stonework of the men's colleges. But having penetrated to the opposite, south-facing side, I found the prospect one of unexpected delight. From this aspect the individual buildings appeared cheerful and distinctive, with many large, white-framed windows looking out onto extensive and secluded grounds. Broad lawns and attractively designed gardens were broken up into separate informal sections by fine old trees, groups of shrubs, and gravel paths, which invited exploration. Beyond were playing fields and tennis courts.

Figure 6.1 Newnham College. (Courtesy of Newnham College Archives)

Heartened, I sought out the office of the Principal, Dame Myra Curtis. I was looking forward to meeting a real live Dame, though I did not like the name of the title, since it reminded me irresistibly of the traditional comic character in a pantomime. She turned out to be a plump and amiable person, with an easy, down-to-earth manner, but I was aware of a strong and determined personality as she looked at me intently through her thick horn-rimmed glasses. She greeted

me warmly, and encouraged me with remarks about Newnham being very glad to have overseas students to broaden
their horizons. Then, having enjoined me to maintain faithfully the College's jealously guarded traditions, she passed me
over to a staff member called Miss Pybus, who, she said, was
to be my Tutor.

Miss H J Pybus was a more serious, but pleasant person,
who spoke with the quick, clipped accent which I soon found
to be a Cambridge characteristic, not always easy for me to
follow. I started to talk to her about the Cavendish and my
aspirations in nuclear physics, but quickly discovered that she
was a historian, with little interest in science, and that her
function as Tutor concerned my personal rather than my
academic standards: the technicalities and rules of behaviour
applying to those privileged to be members of Newnham.

I would be a member of Sidgwick Hall, for which she was
Tutor, Miss Pybus explained. But as Newnham was not large
enough to provide rooms for all students, post-graduates were
housed in nearby licensed digs, which had been carefully
inspected and approved by the College. She felt sure that I
would like the accommodation that I had been allocated in the
home of Mrs Rich, in Grange Road. All my meals would be
taken in College, and to it I must relinquish my ration book.
This was obviously necessary but left me feeling uncomfort-
ably bereft since rationing was very severe in 1946—worse
even than during the war, I was told. I had no means of
buying any extras like bread, butter, jam, biscuits or sweets,
and was at a disadvantage compared with English students
whose families sent them occasional supplements. The College
did give us, once a week, a small section of a loaf, and perhaps
a knob of butter, but I often craved for additions to the frugal
college meals. Power was restricted too, and I was allowed
only one bath a week, in College. All the baths had black
lines painted around the tub at a height of about three inches,
to indicate how much water we were allowed to use.

Miss Pybus went on to explain that, as a student new to the
University, I would be required by the regulations to "keep
nights" during full term for my first two years. Mrs Rich
would have to check me in each night and to provide the
College with a certificate to that effect once a term; I must
also abide strictly by the College rule of being in before
11 pm. Miss Pybus then proceeded to underline the importance that the College attached to the maintenance of the
proprieties by Newnham students: stong moral principles and

exemplary conduct at all times were important College traditions which must be ardently upheld.

I was somewhat surprised at the degree of emphasis which Dame Myra, and now Miss Pybus, had placed on the importance of personal behaviour, and supposed that I might be regarded as being at a double disadvantage. First I had not had my undergraduate training in College, which evidently could not be bettered. Secondly, since I had come from Australia, my credentials were beyond the reach of local assessment, and might possibly not meet Newnham standards.

It was a little time before I began to appreciate the reasons behind the College's apparent preoccupation with moral principles. I then realised that there had probably been no particular concern about me personally. Rather, their attitude was related to Newnham's history and the struggles of women in Cambridge to gain a foothold in the intensely male-oriented University. For several decades after the establishment of Newnham and Girton, women were permitted, on sufferance only, to attend University lectures and to sit the Tripos exam, but they were not admitted to degrees. Bitter campaigns were fought in an effort to improve their status, but the male opposition was obdurate. The women's colleges, acutely conscious of the precariousness of their existence, took great pains to ensure that their critics could not complain about their students' behaviour, and laid down very strict rules of conduct.

It was not until the early 1920s, amidst another round of fierce argument, that the women succeeded in obtaining some further concessions: they were formally admitted to lectures and laboratories in the University and could be given, by diploma, the title of a degree. But they failed to achieve their goal of admission to membership of the University, with the many rights which this would have entailed.

This was still the situation when I arrived in 1946, but, unbeknown to me, the women's case for full membership of the University was about to be examined once again. In retrospect I can well imagine Myra Curtis and the Mistress of Girton both on tenterhooks for fear of any incident on which hardline opponents of the scheme could seize in order to overthrow it; hence the emphasis on women's standards. There were indeed some vociferous protests from anti-female stalwarts. One particular don, referring to the University Proctor and his two attendants, known as Bulldogs, whose function it was to walk the streets at night and to discipline undergraduates, was reported as saying: 'The next thing we'll be having, if we're not careful, will be a female Proctor, followed by a couple of Bitches.'

In the event, the motion proposing the admission of women to full membership was passed in December 1947. It was quite a coincidence that I should be in residence on this momentous occasion in the history of Newnham. In October 1948 the first woman's degree at Cambridge was conferred in the Senate House, on the Queen (now Queen Elizabeth, the Queen Mother)—an Honorary Degree of Doctor of Laws. Afterwards she visited the women's colleges. At Newnham a few "representative" students were presented to her, including myself as an Overseas Student. Since, in common with most Australians, I greatly admired and respected the Royal Family, this gave me quite a fillip.

My digs, comprising a study and a little bedroom, were in an immaculate modern house standing back from Grange

Figure 6.2 The Queen (now Queen Elizabeth the Queen Mother) at Newnham in 1948, on the occasion of the conferring on her of the first woman's degree in Cambridge. She is speaking to the author with the Principal of Newnham, Dame Myra Curtis, standing by.

Road, about 200 metres from the nearest entrance to Newn-ham, and occupied by a don, his wife, and their small daughter, plus one other student. Mrs Rich greeted me in a pleasant and kindly way, though I sensed in her a natural reserve and fastidiousness. I suspect that she had not previously had students in the house, and was not sure that she liked the inconvenience and responsibility. But I was comfortable enough, apart from the problem of laundry. Mrs Rich made it clear that my washing was not to be visible to the neighbourhood, and could certainly not be hung up out of doors. I could put it in the loft, she said. Access to this was by way of a ladder which could be pulled down to the landing; a rope was strung across from the rafters for hanging the clothes. But there was no flooring, and I had to step gingerly from one joist to the next, apprehensive of accidentally putting my foot on the plasterboard in between and having it break through the ceiling of the room below. Apparently Marjorie, my fellow student, did not appreciate the hazard of stepping on the plaster, and one evening, coming back as it happened from seeing a war film with sequences of bombs dropping on houses, I was stunned to find a gaping hole in the ceiling of my room and a heap of rubble on the floor. But Mr and Mrs Rich were in fact remarkably philosophical about that disaster.

Marjorie's successor, Anne, also had laundry problems. One day, in a hurry, she washed out a pair of what were then known as knickers and hung them in her room, near the window. Later, an irate and discomfited Mrs Rich reported that one of the neighbours, who lived in a house beyond the Rich's garden, had telephoned to complain that the knickers were visible through the window. Anne, a somewhat lackadaisical, but lovable person, of whom I became quite fond, moved out soon after that episode; and I felt homesick for my more easy-going way of life in Australia.

To return to my first weekend in Cambridge: my most pressing need was to purchase a bicycle, and to learn to ride it as quickly as possible. Unfortunately the mastering of this art had passed me by as a child, Sydney being too hilly to encourage much use of cycles. But in Cambridge this was the essential mode of transport. On the Sunday morning I wheeled my new acquisition out to a quiet country track, accompanied by Ron, who already had the skill and was prepared to advise and help me. It was a painful process at first, involving close encounters with several bramble bushes, but suddenly, with shouts of encouragement from Ron, I began to make some headway in a wobbly fashion. My confidence and performance improved steadily, until I was overcome by exhaustion.

I slept soundly after my unusual exertions. Then it was Monday morning, 7 October, the day, anticipated for so long, when I was to present myself at the Cavendish Laboratory.

Chapter 7

Paradise Lost

My approach to the Cavendish Laboratory was by Botolph Lane, a tiny, beckoning passage leading off King's Parade just along from Fitzbillys, the well-known bread and cake shop from which, very occasionally, I was to be lucky enough to relieve a little the rigours of rationing by the capture of a cake. On my left was the serene old churchyard of St. Botolph's, and on the right a row of picturesque little shops and houses which had evidently been there long before the Cavendish was thought of.

With rising excitement I turned left into Free School Lane to catch my first glimpse of the famous Laboratory. Standing opposite the back of Corpus Christi College, its appearance was satisfyingly dignified and venerable. The solid stone façade, with its mullioned windows, tall dormers, and little castellated tower, was blackened by successive decades of the products of Cambridge fireplaces and looked older than its seventy-odd years. An outstanding feature of the building was the arched entrance, ornamented above with stone sculptures and flanked by massive oak gates, deeply carved, and dark with age. These, I had learned, were ceremoniously closed and locked by the porter each evening, in accordance with a tradition which had been strongly upheld by Rutherford; he believed that experimental work should not be carried out at night: the participants should go home and think about what they had done and what it might mean, before moving on to the next step. However, I noticed a small door let into one of the big gates, through which access might be obtained by anyone privileged to possess a key.

Beyond the entrance was a spacious, cobbled, oak-roofed, passage. On the right lay the Porter's Lodge; on the left, a few steps led up to the entrance to the Old Cavendish Laboratory itself. So here was I, a callow little antipodean, standing at the threshold of one of the most famous, perhaps the most famous, physics laboratory in the world.

I recalled again, with a sense of exhilaration, what I had learned of Lord Rutherford and the Cavendish, particularly from Pawsey, at the Radiophysics Lab, and from Dr Briggs, in the moving talk I heard from him as an undergraduate. I imagined the exciting atmosphere that must have prevailed here: the inspiration of the great man's boisterous enthusiasm, and his brilliant interpretations of the experiments through which the nature of the atomic nucleus had been unfolded. It had already seemed a remarkable enough fact that everything in the physical world was constructed out of different arrangements of not more than 92 different types of atoms (the chemical elements). But Rutherford and his team showed that the basic structure of matter was even simpler than this. All atoms consisted solely of protons and neutrons in a tiny central nucleus, surrounded by electrons in a planetary-like cloud represented by Bohr's quantum theory, the number of protons determining the chemical identity of individual atoms. Thus Nature had made the whole universe, in all its glory: a butterfly or a mountain, a star or a human being, from the varied dispositions of just three "elementary particles". The simplicity of this picture has been modified by subsequent researches at a deeper level, but the beauty and essential truth of the Rutherford–Bohr atom have not been transcended.

The main Nuclear Physics Laboratory was now, I knew, in a more modern building, located inside the big courtyard. The Old Cavendish had largely been taken over by the Radiowave Research Group, as well as providing accommodation for undergraduate practical classes. But Rutherford's old lab in the Tower had remained untouched, mainly because of its high level of radioactivity. Later I visited this room and saw, amidst the few dusty relicts, the signs of its former occupancy: old flasks and bottles with yellowed labels saying 'Do not touch. E.R.', and one note which read, cryptically, 'The radium source has been moved to a safe place'. The Tower Room, with its windows commanding an extensive view over neighbouring roof-tops, had been used regularly during the war by a team of voluntary fire-watchers looking out for incendiary bombs. They must have received radiation doses well above presently-accepted levels. But radiation hazards were not fully appreciated in those days.

The fame of the Cavendish Laboratory did not rest solely on the Rutherford era. The first Cavendish Professor, James Clerk Maxwell, appointed in 1871, was one of the most brilliant theoretical physicists of all time. He conceived the idea of electromagnetic waves propagating in space, thus presaging radio communication as we know it today, and he established the theoretical basis for this phenomenon. After Maxwell's death in 1879, the Cavendish Chair was occupied for five years by Lord Rayleigh. He, it is said, started the "sealing wax and string" tradition for which the Cavendish became well-known. The principle was still largely being maintained even in the post-war era, as I was soon to discover, with simple components such as tobacco tins and Meccano parts featuring prominently in people's home-made apparatus. I slipped very easily into this way of operating, having learned, during the economic exigencies of my childhood, to exploit the limited materials available to me for making my models and "experiments".

Lord Rayleigh was succeeded in 1884 by Professor J J Thomson, only twenty-eight years old at the time. His most famous achievement was the discovery of the electron, which he showed to be a basic constituent of all atoms, and the carrier of electric current. In 1895, during the Thomson period, the University decided to allow students who had graduated at other Universities to come to Cambridge to take a higher degree. This opened the Cavendish door to overseas students. One of the first of these was Ernest Rutherford, from New Zealand.

J J Thomson's reign continued until 1919. He then handed over the Professorship to Rutherford, who had meantime been at McGill in Canada and then at Manchester University. Now a mature physicist, with an outstanding international reputation, Rutherford plunged into new discoveries almost immediately. In the ensuing couple of decades the Cavendish rose to new heights of fame, which echoed round the world. This was the Cavendish that I now saw myself joining, with an overwhelming sense of privilege. Rutherford was gone, but there must still be the men who had worked with him—carrying on the traditions, blazing new trails in nuclear physics, and inspiring young students like myself.

Looking back, I can see that it was unreasonable of me to suppose the Cavendish of 1946 to be anything like it was in the 1930s. That pre-war era, with Rutherford presiding like a god over the birth of nuclear physics, represented a brilliant peak of scientific endeavour. But Rutherford's untimely death in 1937, followed so soon by the onset of World War II, broke irrevocably the spell of that "golden age". The participants dispersed, most of them taking on prominent roles in wartime research, particularly in radar, and in atomic energy. In 1940 a small group of physicists was formed at the Cavendish, as at some other British University centres, to investigate the possibility that the recently discovered phenomenon of uranium fission could be applied to the making of a bomb. They were impelled by a great sense of urgency: if a bomb were feasible, then Britain must achieve it before Germany, where, indeed, the process of fission had first been established, by Hahn and Strassmann in 1938. Later the Cambridge group broke up, the participants transferring mainly to Canada and the USA. By the beginning of 1945 the Cavendish was virtually deserted.

After the war most of the scientists who had been in Rutherford's group at some time during the 1930s returned home to take up posts in various research centres in Britain. Only very few of them came back to the Cavendish to restart basic nuclear physics research there. They were of the younger generation: men who had just completed their Ph D research before the war broke out. They found themselves very hard pressed: by the urgent necessity of setting up fresh undergraduate courses, lectures, and practical classes, and by the flood of research students who had been accumulating during the war years. Furthermore, the Nuclear Physics Department found itself without a leader. The Cavendish Chair had passed to Sir Lawrence Bragg, the distinguished crystallographer, who was busy setting up an x-ray crystallographic research centre; he had little time to concern himself with nuclear physics. Dr Cockcroft, who had been appointed Jacksonian Professor of Experimental Physics in the Cavendish just before the war, had almost immediately been swept up into wartime work, in radar and then in the atomic energy project, and never in fact occupied the Chair. His appointment in 1945 as the first Director of the Atomic Energy Research Establishment at Harwell left a vacancy.

Pending the appointment of a new Professor, Bragg and his deputy, Mr J A Ratcliffe, who was leading the radiowave research in the Old Cavendish building, agreed that the Nuclear Physics Department should be put into the charge of Mr E S Shire, recently returned from his wartime work, though his experience in nuclear physics was limited. In the Department's chaotic state in that immediate post-war period, he was distinctly over-stretched and could do little but attempt to hold the material threads together on a day-to-day administrative basis.

Shire was nominated Supervisor to all new research students—it being a statutory requirement for the PhD course that each student had an officially appointed Supervisor. But he passed them over, as soon as they appeared, to the other nuclear physics staff members. Most of the British-based students arrived in late 1945 or the summer of 1946. By the time I appeared, as instructed by my College, in October 1946, the Nuclear Physics Department was positively bulging at the seams, with the staff and the facilities saturated. There was no lack of enthusiasm or talent amongst the young staff and more-than-average mature students, but strong leadership was missing, and they were struggling as individuals rather than as a team. The spirit of Rutherford hovered as a nostalgic memory rather than as a driving force.

However I was quite unaware of all these complexities as I stepped forth expectantly towards the new Cavendish building.

The Austin Wing was a large, solid, brick building, plain and functional in appearance. Financed by Lord Austin, the car manufacturer, it had been planned in Rutherford's time, but was barely completed before the outbreak of World War II. Almost immediately it was taken over by the Navy, so that it had only come into use for its original purpose in 1945. The entrance felt rather cold and inhospitable after the mellow warmth of the Old Cavendish. The vestibule was large and tall-ceilinged, hardly relieved in its bareness by solid parquet flooring. On the left I noted the storeroom, visible behind a counter, from which orders would be dispensed, no doubt grudgingly, by a wary storekeeper. On the right were a wide staircase and the entrance to the main workshop. In front of me a long, darkish corridor led down the length of the building. It looked shabby, due probably to the effects of its wartime occupancy, and the high ceiling made it seem narrow. A number of nameless doors on either side gave entrance to small laboratories.

There was little sign of life as I walked uncertainly along the passage, until someone suddenly came rushing out at the far end of the corridor and disappeared again, shouting 'Help! Someone's just been electrocuted.' I never heard anything further about this incident, so presume that it was a false alarm, designed perhaps by fate merely to unnerve me.

The first floor had much the same appearance as the ground floor, but the second floor seemed brighter and more cheerful. The corridor, with a lower ceiling, was well lit by daylight flowing in from a spacious alcove situated half way along. This had a large window looking across to the rooftops of the Old Cavendish, and was furnished with a few comfortable chairs and a coffee table. Around the walls some old, no doubt historic, pieces of apparatus were arranged in cupboards, and there were some interesting photographs of Cavendish staff and students of former years. I noted the office of the Cavendish Professor, Sir Lawrence Bragg, just across the corridor.

In due course I located the office of Mr E S Shire, who had been named as my Supervisor. I knocked timidly and entered. Shire turned out to be a small, prematurely grey-haired man, with a harassed look. He talked rapidly, in staccato bursts.

'In charge of administration for the whole Nuclear Physics Department... many new research students... I've been appointed Supervisor to all the incoming students... only nominal... they are distributed amongst the staff members... Dr Dunworth will be your Supervisor... He isn't here at present... Will be coming soon... will see you then, and will look after you and give you a project... Meantime you can get your bearings... talk to people... see what's going on... Can't talk to you myself... much too busy... will ask Dr Devons to show you round.'

He was a bit like Lewis Carroll's White Rabbit. I would not have been unduly surprised if he had suddenly pulled a watch out of his waistcoat pocket and then gone scuttling off, muttering, 'Oh my ears and whiskers! How late it's getting!' In fact it was I who found myself scuttling off, wondering what to do next.

Dr Sam Devons duly showed me round. He was a distinctly formidable person, in his early thirties I guessed, brusque and sharp-tongued, but with an evidently quick and powerful intellect He seemed bent on deflating me: not so much for my abysmal ignorance of nuclear physics, which he seemed to take for granted; nor even, I think, for my timid attitude of wide-eyed wonder; possibly because of my sex; but most likely out of sheer devilment, mixed perhaps with a touch of nostalgia for the Rutherford period, of which he had had brief experience as a research student. At any rate, he took pains to impress upon me that it was no use my expecting nuclear physics research to be the exciting and apparently glamorous affair that it had seemed in the past. The basic discoveries had now been made. The nucleus consisted of protons and neutrons, we knew. But what of the internal structure of the nucleus? How were we to begin finding out about that? Progress would be painful, and intellectually demanding. New types of experiments must be devised, new techniques invented, new theories evolved. Worthwhile results would not come easily. Far too many people were flocking into the field. What did they all think they were going to do? It would be a case of survival of the fittest. And what did I know about quantum mechanics? No one was going to get far without a proper understanding of the theory... On these themes he carried on relentlessly.

However, his cautionary words affected me less than he may have expected. In my humble state of mind I was not anticipating dramatic personal achievements, or an easy ride. Against all the odds I had got myself to the Cavendish and I was determined to make the most I could out of it: to lap up the wisdom of the first-class minds assembled there; to find a project, with the guidance of a Supervisor, and to slog away at it as hard as I could. Hopefully I might achieve some final results which would constitute a small cog in the wheel of progress in nuclear physics.

The "guided tour" that Devons gave me, after he had exhausted his dissuasive prologue, was in fact a good one. The most vivid memory I have of it is my first sight of the two high-voltage machines. They stood in awesome immensity in a huge hall, which must have been about fifty feet high and perhaps even more in length. They looked like imaginative ultra-modern sculptures. Each had two pairs of slender brown insulating columns, carrying many circular fins and large, flattened, regularly spaced doughnuts of silvery-grey metal. The doughnuts were linked by a zig-zag lattice which rose to a huge bun-shaped metal terminal, somewhere up near the roof, surmounting and dominating the whole structure. The overall appearance of the design, in its elegant proportions, was beautiful—the more so, in fact, for being entirely functional— and it made a dramatic and unforgettable impact on me.

As I stood gaping in awe, Devons was discoursing about the basic design. I rallied my powers of concentration. These machines were the successors to the original high-tension set built by Cockcroft and Walton for their famous 1932 experiment. Made by the Philips Company, they were on a much bigger scale, but followed the same design principle: a voltage-multiplying system which produced a high voltage at the top terminal. The smaller of the two machines, known as HT1, had been bought in 1937 and gave important service during the war. It could generate a positive voltage of up to one million volts. The larger, more recently acquired machine, referred to as HT2, was able to reach a voltage of 1.5 million volts.

The high voltage from each machine was fed via a long fat tube to the top of its associated accelerator, a tall, insulated, high-vacuum tube, surmounted by an ion source (a hydrogen gas discharge), from which a beam of protons could be extracted for acceleration. The accelerators stood on the flat roofs of their respective control rooms, built into either end of the HT hall. The HT1 accelerator was under the jurisdiction of Dr Burcham. Devons himself was responsible for HT2.

Devons went on to explain that the external design of the machines, with their smooth rounded surfaces and their huge size, was dictated by the requirement that electrical break-

Figure 7.1 The Cavendish HT1 Cockcroft–Walton type accelerator. Part of the HT2 machine is visible on the left. (Courtesy of the Cavendish Laboratory)

downs must be avoided. Sharp points, or even specks of metallic dust, would produce very strong local electric fields causing ionization of the surrounding air, and long sparks, like lightning flashes, could occur to nearby points at a lower voltage, or to the walls or roof of the building, if these were too close.

'If you happened to stray into the hall while one of the machines was on,' Devons remarked gleefully, 'your hair would literally stand on end as you got charged up in the electric field. This happened once to the Professor's secretary. Not knowing that HT2 was operating, she tried to use the hall as a short cut. Feeling a sudden tingling sensation, she just stood rooted to the spot and screamed and screamed.' He cackled with laughter at the recollection.

We returned to the Austin Wing via the HT1 control room. Though of normal ceiling height, it had looked dwarfed when viewed from the big HT hall. But inside it seemed quite spacious. The beam pipe from the accelerator could be seen coming through the ceiling and disappearing into a large electromagnet, where the proton beam could be bent through a right angle to emerge horizontally, ready for bombarding a target in a chamber at the end of the beam pipe. The room was busy with people making preparations for the next run and the starting up of the machine, which had been shut down temporarily while Devons and I were in the HT hall. I felt a strong urge to be a part of this activity, and hoped that I would have the chance of running on the HT1 accelerator in due course.

Later that day, when the Cavendish staff and students were gathered for afternoon tea in the Austin Wing tea-room, I saw Mr Shire approach Devons.

'Did you talk to Miss Freeman?' I heard him ask.

'Oh yes,' replied Devons, offhandedly, 'I debunked the atom for her.'

If Shire was reminiscent of the White Rabbit, then there was something of the Queen of Hearts in Devons: 'Off with his head! . . . Off with her head!'

I was told some time later that one or two of Devons' less resilient students, who did not meet his standards and could not face up to his ruthless frontal attacks, actually packed up and left. But those who did survive developed great respect for his intellectual and practical abilities, his extensive knowledge, and his shrewd judgement, and felt that he had taught them a great deal.

Devons was particularly skilled in theoretical physics and in advanced mathematical techniques, including the then unfamiliar methods of group theory. One of his research students told me once how he brought him a calculation that he had made, which extended to several pages. Devons glanced at it and said: 'I could do this in three lines, using group theory.' And indeed this is just what he did, with smug satisfaction, to the wonder and mortification of the student.

For the next few days I wandered around the Austin Wing trying to find out what people were doing. The most responsive and sympathetic person I came across was a New Zealander called Alex Baxter. A solidly built, hearty, enthusiastic man, probably slightly older than me, he had an air of self-confidence which gave me the impression that he must be a member of staff, although it turned out that he had arrived at the end of 1945 as a PhD student. He was occupying part of the Techniques Laboratory—a large room where the vacuum systems, targets, and special devices for the Nuclear Physics Department were made. The technician in charge of this lab regarded me with an air of suspicion and resignation which he apparently extended to all new research students, irrespective of their sex. His job was to serve the needs of the Department and to help students who wanted to make their various pieces of apparatus. As I later discovered, he was remarkably tolerant of students and when one of them did something inconveniently stupid or damaging he would merely remark 'Some mothers will 'ave 'em.'

Baxter, however, knew what he was doing, and was clearly a welcome tenant. He was interested, he told me, in making a new type of detector for the short-range alpha particles from nuclear disintegrations which he was going to study using the HT1 set. Conventional gas-filled detectors, such as ionization chambers and Geiger counters, were slow in response, and had to be sealed with a thin window which absorbed some of the energy of the alpha particles before they entered the detector. What was needed, Baxter explained, was a high-vacuum detector which could be coupled directly to the vacuum system in which the nuclear reaction particles were being produced. At present he was experimenting with the design and construction of an electron multiplier, a device, just being developed in the United States, which contained a series of electrodes made of special secondary-electron emitting material.

Designing and making such an object seemed to me a difficult undertaking for a research student, but it was evident that Baxter was already quite experienced in valve manufacture and laboratory techniques. He talked to me at length about problems involved in his work, showing me trial components which he had already made for his electron multiplier. As well as being technically competent, he also had considerable powers of verbal exposition: I subsequently saw him in action with visitors to the Cavendish, including some of high academic standing. They were happy to be shown around by Baxter and referred to him deferentially as Dr Baxter, a form of address which he did not bother to disown. I was also to discover that he was very willing to advise and help other research students in their technical problems. I suspect that many experimental systems of that period bore something of the Baxter stamp upon them.

I was very grateful to Baxter for the kindness and understanding he showed me that day when I first appeared uncertainly on his doorstep. I told him of my present want of a Supervisor, which, combined with my lack of knowledge of practical nuclear physics, left me more or less stranded. 'I've never even used a Geiger counter,' I confessed, knowing that this was a standard device for detecting particles emitted in radioactive decay, through the ionization they produced in the gas contained in the counter. 'Well, why don't you have a go with one now?' said Baxter cheerfully. From the stock of bits and pieces of apparatus he had around him, he produced a Geiger counter—a short metal cylinder sealed at one end with a thin mica window. He set it up with a suitable voltage supply, amplifier, and recorder, and provided me with a radioactive source deposited at the end of a metal rod. Then he left me to it.

I had quite a field day exploring the features of that radioactive source. Sitting at a corner of one of Baxter's benches, I was able to repeat in a simple way some of Rutherford's pioneering radioactivity experiments. Holding my source close to the window of the counter, I got a high counting rate. As I started to draw it back, the counting rate remained fairly steady until quite suddenly it dropped dramatically. The distance at which this happened was a few centimetres of air. Thus, with allowance for the residual stopping power of the counter window, I had roughly measured the range in air of the alpha particles from my source.

Some counts still persisted when the source was withdrawn further. These, I realised, must be due to electrons emitted in the radioactive process known as beta decay. A thin slab of lead, with which Baxter had thoughtfully provided me, inserted between the source and the counter, stopped the electrons, as was to be expected since lead is very much denser than air. But there were still a few counts being registered in the counter. These, I told myself, must be due to gamma rays—quanta of very high frequency electromagnetic radiation—emitted from excited states of daughter nuclei formed after alpha or beta decay. They could be reduced only by the insertion in their path of a thick lead brick.

As I followed through the various procedures, my bits of textbook knowledge started to fit together, like the pieces of a jig-saw. A new, almost tangible, picture of the radioactive radiations emerged, and I could visualize individual nuclei in my radioactive source spontaneously spitting out particles and transmuting themselves into different elements. My recollection of that experiment is printed deeply in my memory: a moment of sudden comprehension, generating a surge of intellectual excitement; a moment of the kind that has been for me one of the great rewards of studying physics. I felt grateful to Alex Baxter for having given me the opportunity and freedom to pursue that simple experiment without making me feel small and ignorant.

Baxter was in fact to play an even more significant role in my Cavendish work, as will be related in a subsequent chapter.

After some days I went back to Mr Shire to ask about my Supervisor. No, Dr Dunworth had not appeared again yet, he said. Meantime it might be best for me to join in with one of his two students, who were working in another building. 'Go and see John Hill,' he suggested.

I found John Hill occupying a fusty laboratory room on the first floor of what was known as the Old Anatomy Building, on Corn Exchange Street. It lay across the yard from the Austin Wing, beyond a low building through the grimy windows of which could be seen some dejected-looking relicts belonging to the Zoological museum. John greeted me with comforting affability. He was a slight, fresh-faced young man, with a cheerful, easy manner and a ready sense of humour; his conversation was enlivened by typical expressions of Air Force origin, like "wizzo" and "good show"; he explained that he had been in the Service during the war as a radar engineer.

'I wanted to be in airborne radar,' he said. 'But at the end of our initial training we were told that those with surnames beginning with letters A to M were to be ground-based, and N to Z were to be airborne; so that was that. 'Anyway,' he added reflectively, 'if I had been in the second half of the alphabet I probably wouldn't be here now... Many of them never came back.'

After the war John decided that he would like to get a Ph D degree and he started as a research student at the Cavendish in February 1946. He told me that he had been allocated to J V Dunworth who was in the Cavendish at that time, having just returned from working in the Atomic Energy Project in Canada. Dunworth's supervision consisted mainly in handing John, and his other student, Les Shepherd, a copy of his (Dunworth's) Ph D thesis and suggesting that they carry on from where he had left off. They had seen very little of Dunworth after that, and recently they hadn't seen him at all.

The project which John had thus acquired was that of measuring the half-life of the radioactive nucleus then known as thorium C'. This is a member of the radioactive series starting with the naturally occurring element thorium. Successive decays, either by alpha- or beta-particle emission, occur through a chain of elements until a stable product is reached. Each element in the chain has a characteristic half-life (the time for half of an initial number of nuclei of that element to have decayed). The half-lives of individual members of the radioactive chain vary enormously: from ten thousand million years for thorium itself (a measure of the age of the Earth), to an extremely short value for thorium C', which Dunworth, introducing an original coincidence method, had found to be a fraction of a microsecond (a millionth of a second). But there were some deficiencies in his measurements, and it was John's task to try to improve on them. For this he was able to draw on the experience of microsecond pulse techniques that he had gained in his wartime radar work.

He took considerable trouble to explain to me how he was going about his experiment, and he also introduced me to his fellow student, Les Shepherd, who was working in an adjoining lab. Les, also very friendly, was a more serious and reserved character. I did not feel as completely at ease with him as I did with John, particularly when I learned that he was a founder member of a society called the Cambridge Interplanetary Society, which was carrying out detailed theoretical investigations into ways and means of propelling rockets and other vehicles into interplanetary space. In the year 1946 such enterprises sounded so impracticable that I (and others at the time) could not take them very seriously. But we were of course all confounded by subsequent events. Shepherd and his colleagues were then proved to have shown remarkable prescience.

Shepherd was studying the electrons emitted in the beta decay of thorium C, the progenitor of thorium C' in the radioactive chain; John was measuring the very short time intervals between the formation of thorium C' nuclei, by beta-decay of the thorium C parent, and their subsequent decay by alpha-particle emission. For detecting the particles, John was using a Geiger counter. He explained to me that one of the problems occupying him at the moment was that of allowing for the variable delay in the firing of the Geiger counter after it had received an ionizing particle. It seemed that this might significantly affect his time measurements. Some experiments should be done to investigate this, he said.

'If you are at a loose end until Dunworth turns up, why don't you do some of these experiments?' suggested John, enthusiastically. 'I think the best way of detecting these delays is to exaggerate the effect by making a special Geiger counter, much longer than the normal one. Start the ionizing discharge with a radioactive source at one end of the counter and find how long it takes for a pulse to be received at the other end. Then we can deduce what the delay will be in the short counter; and Bob's your uncle!'

Thankful that Baxter had given me the opportunity of becoming familiar with the use of a Geiger counter, I said that John's suggestion sounded good to me. I was itching to get my teeth into some sort of work: in my state of increasing frustration even a temporary project would be better than having nothing to do. But then I looked at John's electronic equipment spread along the bench. How was I to cope with designing and setting up a system like this? My knowledge of klystrons and magnetrons wasn't of much use to me in this kind of work.

'Not to worry,' said John cheerfully. 'I can help you design the counter. And all the electronic equipment you'll need for doing the measurements is here.' He swept his hand vaguely in the direction of his homemade boxes and recorders. 'I can show you how to set it up. After all, these experiments are going to be of value to me, so the least I can do is to collaborate with you.'

What a boon John Hill was to me, with his infectious good humour and ready help, during my critical early Cavendish period! And I soon learned that, behind his apparently care-free, easy-going manner, there was a sharp, critical mind. He was in fact to have an impressive career, ultimately reaching the position of Chairman of the United Kingdom Atomic Energy Authority, an achievement which surprised and pleased me when I heard of it. It was encouraging to know that someone as gentle and modest as I had known John to be was able to make his way in a tough, competitive sphere of work. John was also a naturally good teacher, and I gained some valuable knowledge about pulsed circuitry and coincidence methods while I was working with him.

My immediate job, however, was to make a long Geiger counter. A normal counter consisted of a metal cylinder about two centimetres in diameter and five to ten centimetres long. My counter was to be a metre long, with short insulated pieces at each end. First I had to cut and finish the tube sections on a lathe. I had never had the opportunity of learning how to use a lathe, since in the intense wartime days at the Radiophysics Lab the scientists had not been allowed near the workshop. Well, I said to myself firmly, now I had the chance. I had been told that there was a lathe available for student use in the Austin Wing workshop. So there I repaired, in some trepidation, clutching the length of copper tubing which I had captured from Stores.

The foreman, with surprising willingness, agreed to introduce me to the operation of the lathe. Patiently he explained its parts and functions. On the left-hand side, the object to be worked was gripped in the jaws of a chuck, which could be rotated rapidly by a motor. On the right, a cutting tool could be moved continuously by very fine hand adjustments, either along or across the axis of the machine, until it touched the rotating object at the required point and started to pare the metal away. To demonstrate the operation, the foreman took from a scrap-box a suitable brass cylinder, set it between the three massive jaws of the chuck, and clamped it tightly using a special square-section metal key which he inserted into a square hole at the side of one of the jaws. Then he selected an appropriate hardened steel cutting tool and clamped this to the tool post. Now he was ready to start. He switched on the motor. The brass cylinder began to spin. Moving the cutting edge up to it slowly, he started the turning process, adjusting the tool position as metal was shaved off in thin curling strips. It looked simple enough. Then, taking the cutting tool back a little, the foreman switched off the motor and left me to try it for myself.

Timidly I practised for a while on the dummy cylinder, gradually getting the feel of the control of the cutting tool. It seemed to work quite well. So then I proceeded to the real job. Unlocking the chuck jaws, I removed the practice piece, inserted my copper cylinder, tightened the jaws again, adjusted the position of the cutting tool. All seemed set to go. I switched on the motor and the chuck revved up.

Suddenly, after a second or so, there was an enormous bang, like a rifle shot, followed by a small metallic tinkle. Quickly I switched off the motor, wondering what on earth had happened. Then, as I looked in the direction from which the sound seemed to have come, I realised with horror what I had done. After I had tightened the chuck jaws with the locking key, I HAD FORGOTTEN TO REMOVE IT. As the rotating jaws had been gathering speed, the key, quite a massive object, had gone round with them a few times and then had been flung out at violent speed to hit the opposite wall. I broke out into a hot sweat as I contemplated what might have happened if I, or anyone else, had been standing in the path of the lethal missile. Furtively I looked around.

The whole workshop seemed to be hushed. Slowly, agonizingly slowly, the foreman started to walk down the length of the shop towards me, his face expressionless. While thanking fate that I had not caused a serious accident, I waited for the abuse, the sarcasm, the 'this is the last time I'll allow a woman in my workshop', the ignominy of dismissal . . . .

Now he was facing me squarely. He spoke at last.

'Well I don't think you'll make that mistake again.'

That was all. I couldn't believe it. Maybe, with the numerous students he had already had to contend with, this was not the first time such a thing had happened? Whatever the reason for his leniency, I was reprieved.

So, when I had recovered from the shock, which had certainly taught me an unforgettable lesson, I was able to proceed with making the parts for my counter. With John Hill's advice, and the aid of the technician in the Techniques Lab, the counter was in due course assembled, the glass-to-metal seals were made, and the gas filling was introduced. There were "bugs to be ironed out", as the saying goes: newly made apparatus always seems to have some faults to be corrected. But at last I had the satisfaction of seeing the counter behaving as a Geiger counter was supposed to behave, and I was ready to start the measurements.

My experiment worked out well. I was able to measure the velocity of the discharge along the counter for different voltages and gas fillings. The results confirmed what John had expected from his earlier work: the delays in the firing of his counter would vary by an amount comparable with the half-life he was trying to measure. Thus he had to find a method which would get round this problem. This he finally achieved, and was able to measure the half-life of thorium C' with an accuracy ten times better than that of Dunworth's result.

As it happened, the interest in my work extended beyond that of John's thesis. At the time when I was doing the experiment, a very bright young man who had just completed his PhD thesis, and had become a junior member of staff in the Nuclear Physics Department, was writing a book on the theory of ionization chambers and Geiger counters. His name was Denys Wilkinson—a name which was to become famous in British and international nuclear physics circles. He was including in his book a section on the processes involved in the spread of the discharge along a Geiger counter, and had derived a theoretical expression for the velocity of propagation of the discharge. So my experimental results were opportune. They satisfactorily confirmed his formula, and Wilkinson asked me politely if he might include them in his book. I suppose I might have been able to write up my experiment in a publication of my own, but I had no idea how to go about this, so I was glad to have the results perpetuated by Wilkinson. He thanked me courteously for the graphs with which I provided him. 'Now I have the opportunity of discussing the shape of Miss Freeman's curves,' he remarked, giving me my first intimation of his individual brand of facetious humour, which was also to become famous.

The long-counter experiment was good experience for me, but all the time I was working on it I was painfully conscious of the fact that it was not getting me anywhere as far as my PhD thesis project was concerned. There was still no sign of Dr Dunworth. John Hill had not seen him, and had no idea when he might appear again. I began to wonder if Dunworth even knew of my existence. Mr Shire was completely unhelpful. I went and talked to Dr Burcham, on whose HT1 set I really wanted to work. Burcham, a kind and gentle person, was sympathetic, but said, quite firmly, that he couldn't possibly take on an extra research student. He was already responsible for some half-dozen, and there were others who also required running time on the accelerator. It was grossly overloaded by all these demands. There was simply no room for another student. I even tried the fearsome Dr Devons, with the same result. In desperation I then went to Mr Ratcliffe, who was running the Radiowave Radiation Group where I knew Ron Bracewell was progressing happily, to see if, by reason of my previous experience at the Radiophysics Lab in Sydney, he would be willing to take me on. But I got the same story again: he already had too many students, and their facilities were overloaded. The whole Cavendish was in fact saturated. There seemed just no room for me. How was I to acquire an official Supervisor and a project? And even had I devised my own project, how could I obtain access to experimental equipment to work with? My situation was desperate.

I found that I was not alone in this predicament. There were three other overseas students who arrived at the Cavendish to do nuclear physics at the same time as I did: two

Figure 7.2 From left to right: W E Burcham, S Devons, E S Shire, and D H Wilkinson, in the mid 1950s.

Canadians, George Lindsey and Charlie Barnes, and a South African, Godfrey Stafford. They, like me, were shocked at the evident disorganization of the Department, and the lack of leadership; and they too had the feeling of being "left-overs", too late to be accommodated. Without supervision or projects, they found themselves as stranded as I was, and became equally depressed. But this was not much consolation to me, except for eliminating from my mind the thought that there might be sex discrimination in my case.

The exceptional weather added to my miseries. The winter of 1946/47 proved to be the worst of the century. At first the copious snowfalls thrilled me. I had never in my life before seen even a single snowflake. So the appearance of great cotton-wool blobs, floating silently down, transforming the landscape to Christmas-card unreality, excited me so much that I capered about in this wonderland like a child, to the cynical amusement of my hardened English friends. Exaggerated epaulettes of snow formed on my shoulders as I cycled precariously along the rutted streets. Mundane objects like dustbins and letter boxes assumed a new, picturesque form. Trees and bushes were miraculously decorated. Rolling up snow carpets to form huge snowballs was tremendous fun, and I even stimulated some of my Newnham companions to join me in this sport, until I was ticked off, by the Principal of Newnham in person, for damaging her rose-bed, which was buried underneath.

The thickness of snow mounted steadily, and soon everyone was walking between high walls of piled-up snow, where attempts had been made to maintain narrow footpaths and roadways. Cycling was virtually impossible, and cars jazzed about precariously. The temperature dropped steadily, with no prospect of a thaw. Contingents of lorries were sent into the Town Centre, laboriously to cart away some of the now mountainous obstacles to movement. Transport throughout the whole country was disrupted. Railway lines and main roads became blocked. Movement of coal virtually ceased. Stocks at power stations dwindled, and, as the cold continued to intensify, power restrictions had to be imposed. My excitement became seriously tempered by discomfort. Those of us who lived in digs which depended on electricity for heating found our rooms intolerable to stay in for any length of time. Although no effective work could be carried on in the Cavendish because of the power cuts, most of us came into the Lab nevertheless, huddled in disconsolate groups to avail ourselves of the bit of warmth still obtainable from the heating system there.

Occasionally we treated ourselves to a cafe meal we could ill afford, to supplement the ration-limited, semolina-based College diet. I found myself thinking longingly of home, warm Australian sunshine, real food, and the companionship of my mother, who, I realised, must be suffering agonies of disappointment over my Cavendish situation as she read between the lines in my letters.

One day a food parcel, sewn with infinite care into a stout canvas cover, arrived for me from my mother. She had packed into it a number of items appropriate for me to take to the kind family who had offered to house and entertain me during the next vacation period, when I would be required to vacate my digs. Amongst the items was a tin labelled "home-made beef dripping". Could I possibly save this for Easter? The temptation at least to open it was irresistible. It had been beautifully sealed and was in perfect condition, with a mouth-watering smell. I sampled a spoonful. Delicious! Before long I had devoured the lot.

One evening, coming home late to my digs and preparing to go to bed in my frigid room, I discovered that my hot water bottle was still inside my bed, left there from the previous night. Extracting it, and holding it over the washbasin as I uncorked it, I watched with horror as a slow stream of slushy ice filled the basin. This was the last straw. My morale, already seriously undermined by the apparent hopelessness of my situation in the Cavendish, and now penetrated by the cold, plummeted to new depths. In a surge of self-pity and homesickness I wept bitterly.

Chapter 8

Paradise Regained

Eventually, in mid March, came the thaw, followed by the Great Flood. Cambridge was all but submerged in a vast expanse of water. Cycling down Sidgwick Avenue from the slightly elevated ground of Newnham College, I was confronted by a lake stretching across Coe Fen from the Causeway to Silver Street and from the Backs to Queens' Road. Newnham village was cut off, and mallards were paddling about noisily by the traffic lights, seemingly surprised by their suddenly extended territory. Released at last from the vicious clamp of the long, hard winter, students were capering about excitedly, exhilarated by this new nine days' wonder. Long diversions had to be sought to reach the town. At the Mill Lane weir a man was on duty, huddled over a tiny coke fire; I wondered what he could possibly do to control the turbulent course of the swollen river.

Gradually, over the course of a few days, the floods subsided, leaving bedraggled skeins of twigs, mud, and debris across the roads and meadows. Then, suddenly, spring, suppressed for so long by the fierce winter, announced itself in a precipitate outburst of glory. Immaculately white snowdrops and lemon-yellow aconites emerged, and, almost simultaneously, crocuses transformed the dull earth with miraculous stretches of vibrant colour. Walking along the famous King's College avenue which gently curves its way from the Queens' Road gate towards King's Bridge, I was entranced by the pure beauty of these lovely flowers, brilliantly yellow, purple, mauve, and white, beneath the dark, still leafless trees. Somewhere a blackbird was singing in rich fluty cadences, and I saw, clinging to the rough bark of a nearby tree-trunk, my first English wren. He was so busy courting his lady-love, swinging his stubby little tail slowly and rhythmically from side to side in an ecstasy of passion, that he seemed oblivious of my close observation. By the river more crocuses adorned the banks, and a profusion of daffodil buds foretold the next wave of colour. Huge weeping willows, bending down to touch the now placid waters, had, about their long thin branches, an aura of palest green, where the new leaf-buds were about to burst forth. Everywhere Nature seemed to be proclaiming, with exquisite grace, a message of renewal and hope.

Leaning on the parapet of King's slender bridge, I let my mind absorb the kaleidoscope of images, and I relished the emotional delights of springtime in Cambridge. A new sense of purpose and optimism possessed me. I must not let myself be defeated by my set-backs at the Cavendish, I must find a way to get myself established there.

How should I pursue my new resolve? It was no use trying another frontal attack. I must devise a more subtle approach. I had decided that I would like most to work on the HT1 set, the successor to the original Cockcroft and Walton machine. Studies of nuclear reactions with an accelerator seemed to me more exciting than bench experiments like John Hill's; the students who were already working on HT1 appeared to be forging ahead with industry and purpose; and, most important of all, Dr Burcham, the man in charge of the accelerator and of most of the students working on it, was evidently a kind, helpful, and conscientious person. I would like to have him as my Supervisor.

I considered how I might persuade Burcham to take me on, even though he had said that he already had as many students as he could cope with. Well, I thought, the first thing to do is to find out where his personal nuclear physics interests lie. This I might well discover by looking up the papers that he had probably published on his PhD work, just before the war.

In the well-stocked Library of the Austin Wing, I soon found, under the name Burcham, references to two papers published in 1939, one by Burcham and Smith, and the other by Burcham and Devons, no less; I discovered that they were both concerned with the investigation of short-range alpha particles emitted from nuclei under proton bombardment.

'Short-range alpha particles,' I repeated to myself. A bell rang in my memory. That was the phrase that Baxter had used when he mentioned to me, on my first day in the Cavendish, the experiment on HT1 for which he was designing his special detector. My brain started working fast: Baxter had arrived at the Cavendish in 1945—very likely before most of the other new Ph D students had turned up. This might well be why he was working on his own, whereas later arrivals seemed to have been grouped in pairs. It was quite possible that Dr Burcham, confronted with his first student, had suggested that he carry on with the kind of experiment which he, Burcham, had been doing just before the war. So Burcham would be particularly interested in this experiment. But what had Baxter done about it? More than a year later he was still concentrating on making a new type of detector. I began to formulate a plan of campaign.

But first I settled down to absorb as much as I could of Burcham's papers. I could not follow all the nuclear physics discussions, but it was clear what the experimenters had done. They had used the accelerated proton beam from the HT1 set to bombard a fluorine target in an evacuated chamber. Many of the protons bounced off the target like billiard balls, scattering in various directions; but in a few cases a proton produced a nuclear disintegration with the emission of an alpha particle. A thin mica window opposite the fluorine target allowed some of the alpha particles to emerge from the vacuum chamber into the air. The object of the experiment was to determine the energies of these particles by measuring their ranges in air. An ionization chamber, somewhat similar to a Geiger counter, was used as the detector. One group of alpha particles was found to have quite a long range, several centimetres of air, but there were other interesting groups which appeared to have much shorter ranges. In Burcham's earlier experiment these could not be measured because the more prolific scattered protons had a greater range than the alpha particles. However, in the later experiment this problem was overcome by the use of a magnetic field.

The short-range alpha particles and scattered protons emitted from the target were allowed to pass, under vacuum, between the pole-pieces of an electromagnet. In the magnetic field the two kinds of particles followed circular paths of different radii and were thus separated from each other. With the correct choice of magnetic field strength the alpha particles emerged through a thin exit-window in the vacuum box and their ranges in air could then be measured using the ionization chamber. This, I realised, was an important step forward. The technique could be used to study lots of different nuclear reactions. Excitedly I put the journals away and proceeded to the first phase of my plan, which was to talk to Alex Baxter.

I found Alex working away happily in the Techniques Lab. I had visited him a number of times since my first contact with him, and knew that he was still preoccupied with his electron multiplier. He greeted me with his usual affability.

'How are you doing, Alex?' I enquired.

'Fine,' he said cheerfully. 'I've been testing my electron multiplier. It now works well under good vacuum conditions.' He went on to explain that he still had a problem, because exposure to air affected the efficiency of his detector. He would have to find a way of treating the surfaces of the electrodes so that they would survive being let up to air while the instrument was being transferred to the vacuum chamber on the accelerator. 'I'm going down to Baldock tomorrow to discuss the problem with them,' he said.

Alex obviously had good connections with the Services Electronic Research Laboratory at Baldock—a well-known centre for valve research which had been particularly important during the war. He seemed to be a frequent visitor there, and must have been able to give, as well as receive, useful technical advice. But his problem was a serious one, and I wondered if his electron multiplier would ever prove practicable for use on the HT1 set. However, Alex did not appear to be disheartened. Clearly he enjoyed challenges related to valve techniques, and revelled in the kind of work he was doing. He had told me on a previous occasion that he had an ambition to establish a firm in Cambridge specializing in valves and high-vacuum equipment. His objective in obtaining a PhD degree was mainly to further his career in that direction; he was really not interested in academic nuclear physics. So much the better for my plan. I decided that now was the moment to confront him with it.

'Alex,' I began, 'I have a proposition to put to you.' I told him that I had been reading about Dr Burcham's pre-war experiment on short-range alpha particles from proton bombardment, and that I could see how interesting it would be to pursue this further. But there was obviously much work to be done in the setting up and preliminary testing of the experimental equipment. I suggested that if Baxter had a collaborator, he or she, with Burcham's help, could be doing all the preparatory work, using an ionization chamber as the detector for the time being, while Baxter got on with developing his electron multiplier. Then, when this detector was ready, he would be in a position to transfer it immediately to the experimental equipment and to carry straight on with joint nuclear physics experiments.

'You can see what I'm getting at, Alex,' I said, looking at him intently, 'so what do you think? Would you be prepared to accept me as a collaborator?'

Alex was a quick thinker and had shrewd judgement. He must appreciate the advantages to himself of my proposition, I thought. But on the other hand he was a very independent character, and might prefer not to get involved with any collaborator. I waited anxiously for his response.

It did not take him long to make up his mind.

'An excellent idea,' he said emphatically. 'I could soon make up an ionization chamber for you, and the preamplifier and electronic equipment to go with it. Yes, I certainly have plenty of development work of various kinds to do here. As a matter of fact,' he added, with growing enthusiam, 'I'm also very interested in the possibilities of a photoelectric multiplier as an alternative type of detector to the electron multiplier. These photoelectric devices are still at an early stage of development, and are not very satisfactory yet, but I would like to go and talk to the people at the EMI Research Labs and urge them to put more effort into producing this type of tube.

'Yes,' he concluded, 'I think it would work out splendidly if we collaborated as you suggest.'

'That's great,' I said, with enormous relief. 'Now I can go and talk to Dr Burcham, and put the plan to him. I hope he'll approve.'

'Oh, I expect he will,' said Baxter with his usual optimism.

Elated with the success of my plan so far, I went off to tackle Burcham. He was in his office, and greeted me courteously.

'Since I have no Supervisor,' I began, 'I would very much like to have a chat with you.'

'By all means,' he said, smiling reassuringly and sitting back to give me his attention.

'First of all I would like to tell you about my long Geiger counter experiment with John Hill,' I began. I produced the graphs of my results, mentioning the agreement with Denys Wilkinson's theory. He made some helpful comments and said that it was a nice piece of work. He's certainly an encouraging person, I thought.

'Now I have to find another project,' I went on. 'I have been reading about your work on alpha particles from proton bombardment, published in the *Proceedings of the Royal Society*. I was particularly interested in your use of a magnetic field to remove the scattered protons, so that you could study the short-range alpha particles.'

I could see that I had now really caught his interest.

'I would dearly love to take part in the continuation of that experiment,' I said. 'I know it is Baxter's project, but I am also aware that Baxter has become so involved in developing a better type of detector for it that he hasn't even started to set up the experiment itself. Now most of your students seem to be working in pairs, so I am wondering if you would consider the idea of my collaborating with Baxter, under your supervision.' I went on to make the suggestions that I had already made to Alex.

I could hear myself talking quite forcefully. Instinctively I was aware that at this critical juncture, with so much at stake for me, there was no place for my usual, more diffident, approach.

Burcham seemed impressed by what I had to say. However, he was obviously cautious by nature. He admitted that he had been disappointed by Baxter's apparent lack of interest in getting the experiment started, but felt that it was necessary for Baxter to apply himself to the nuclear physics aspects of the work as well as to the technical problems of detectors. I pointed out some of Baxter's considerable achievements as well as his present difficulties with his electron multiplier, and I mentioned also the photoelectric multiplier which Baxter thought might be the ultimate answer to effective alpha-particle detection.

'Under your supervision Alex and I should be able to do some very worthwhile joint experiments when the detector problems are sorted out,' I argued urgently.

Burcham smiled in his slow, gentle way. I think my persistence and the fact that I had studied his papers had gained his respect. At any rate, he finally agreed to my proposal. I was in! What a relief! I felt as if I was suddenly on springs. I could have bounced up and touched the ceiling, or leaped about in ten foot strides.

That evening I poured out the story of my triumph in a long letter to my mother, and sowed the seed of an idea which had just occurred to me in my rejuvenated enthusiasm: 'Why don't you get a year's leave of absence from your school,' I wrote. 'Come over to England, find a temporary teaching post somewhere, and share with me some of the delights of Britain and Europe.' That idea bore fruit, and my mother in due course made the trip to England, to my great pleasure and satisfaction.

Looking back at my previous six months. or so at the Cavendish, from the security of my newly found niche, I realised that they had not been entirely a waste for me. I had gained some useful practical experience (including how to use a lathe!), I had completed a small project which was regarded favourably by two members of staff, and one way and another I had learned some basic nuclear physics.

Some of this was acquired through interactions with other Cavendish research students. John Hill's value to me had of course been tremendous. Most of Dr Burcham's students appeared intimidatingly serious and purposeful, and I felt too diffident to approach them. But one or two were very helpful and informative—in particular Harold Allan, a jovial young man with a round, smiling face and outgoing personality, who had started work on the HT1 set at the end of 1945. He took the trouble to tell me about his experiment, and gave me an idea of what working on the accelerator was like. He had been in naval radar during the war and seemed to me extremely competent. He was to become a well-known cosmic-ray physicist at Imperial College, London.

Working with Allan was the only other woman student in the Nuclear Physics Department, a French girl called Christiane Clavier, whose personality, unlike Harold's, was excitable and at times fiery. She also was a Newnham student, and used, on occasion, to regale us at mealtime with a somewhat sensational account of her experiences in running an experiment on the HT1 set. Coming from France, a stronghold of women physicists—perhaps because of the Marie Curie tradition—she had a sound knowledge of physics, and I found my associations with her informative and stimulating.

But Christiane was deadly in an argument. I once had an altercation with her in the HT1 target room about which of us had priority in the use of the accelerator on the following Saturday morning, a period which was habitually a sort of bonus before the machine and its hard-working operator took a weekend rest. I had difficulty in matching the intensity of the screams with which she was bombarding me, her eyes flashing, her hair flying wild. As I opened up my own vocal chords in self-defence, there was suddenly an ominous cracking sound above our heads. Looking up, we saw that a patch of plaster on the ceiling had started to break away and was about to fall in large, heavy chunks. We dashed simultaneously for the door. Laughing somewhat hysterically, we peered back anxiously into the room. Fortunately no apparatus had been damaged by the falling material, which by then had become a heap of rubble on the floor. The ceiling had probably started to crack up before our shouting match began, and the sudden intensity of high-frequency sound had been the last straw for it. There were no witnesses of this incident, apart from ourselves. Birtwhistle, our invaluable, imperturbable, technician, who was responsible for the accelerator and the target room under Burcham's direction, accepted the situation philosophically and soon had the ceiling restored.

After gaining her PhD degree, Christiane, having married, gave up her professional interests in favour of raising a family. However, she returned to physics some time later, when she joined the staff of the Nuclear Physics Department at Oxford University.

Potential sources of enlightenment in nuclear physics were provided by various lecture courses. But I found them highly theoretical, and mostly above my head. Nevertheless I appreciated the opportunity of seeing and hearing some of the outstanding theoretical physicists in Cambridge. The most famous was Professor P A M Dirac, an acknowledged genius, whose fundamental theories have been of outstanding importance to modern physics. I sat in on the course he gave in 1947 to undergraduates studying for the Mathematical Tripos. Details of his subject matter proved to be largely incomprehensible to me, but I kept on attending the lectures for the sheer fascination of watching the master in action, and gradually I acquired some awareness of the power and originality of his mathematical techniques. His admirably logical expositions were delivered in a quiet, economical style, yet with an air of remoteness, as if he were seeing only an aesthetic, abstract world of his own. On the blackboard Dirac would methodically build up a succession of mathematical expressions, using strange notations which he had himself invented. To represent a new pair of operators (an operator being an object which performs some kind of function, much as the familiar "cross sign" denotes multiplication), he adopted bracket signs, one opening to the right, the other to the left. These he called "bras" and "kets", so that his discourse was liberally sprinkled with these terms, "bra" being pronounced as is the word for the well-known lady's undergarment. But not a flicker of a smile did this evoke from his earnest audience. Even had they thought of the connection, I'm sure no one would have felt an urge to giggle in the mesmeric presence of Dirac.

I looked around at these fresh-faced undergraduates, assiduously copying down the contents of the blackboard, and wondered if they were really able to grasp the details and significance of Dirac's mathematical manipulations. I assumed that most could. The academic standard of the students studying for the Cambridge Tripos was indeed in a class of its own. Fortunately, I reflected, one could carry out experimental physics without having to reach these dizzy heights.

I witnessed an example of Dirac's apparently quite innocent adherence to pure logic a year or so later, when I heard him give a special lecture in the Old Cavendish. It was a formal talk in that there was a Chairman who first introduced the speaker. Dirac then proceeded to fill the blackboard with mathematics, addressing his personal unseen world in his characteristic fashion. When the lecture was over, and Dirac had sat down, the Chairman expressed his thanks to the speaker.

'Professor Dirac,' he then asked, 'would you be prepared to answer questions?'

Dirac nodded and the Chairman looked expectantly at the audience. A young man near the back stood up.

'Professor Dirac,' he said, 'I do not understand your fourth equation down on the left-hand side of the blackboard.'

Dirac sat on impassively, staring into space as if he had not heard. A long silence ensued. The Chairman was becoming embarrassed. Finally he intervened.

'Professor Dirac,' he said diffidently, 'are you prepared to answer that question?'

'Oh,' responded Dirac, suddenly coming to life and looking surprised, 'was that a question? I thought it was just a statement of fact.'

Some people to whom I have related this story have questioned its credibility, but not so long ago I happened to meet someone who was also at that lecture. He vouched for the truth of my account, but wondered if Dirac might have done it deliberately, just for fun.

The most valuable source of enlightenment for me on nuclear physics topics, during my earlier Cavendish period, stemmed from the *Nuclear Nit-Wits Club*. This was the brain-child of George Lindsey, one of the two Canadians who, with the South African, Godfrey Stafford, and myself, constituted the overseas student group—the "colonials" as we were sometimes dubbed. In our early tribulations and sense of neglect, as late-comers to the overstretched Nuclear Physics Department, we tended to join forces, sharing our grievances and exchanging ideas on how to overcome our problems. We were very conscious of being at a disadvantage, compared with the English students, in not having had a proper course in nuclear physics in our undergraduate days. The Cambridge graduates in particular had obviously had excellent teaching, and some of them had gained further experience in Canada, in the Atomic Energy Project which had flourished there during the latter part of the war, under Dr Cockcroft's direction.

'What we need to do,' said George, 'is to have regular get-togethers to discuss aspects of nuclear physics which we do not understand, and to work at trying to inform ourselves. Perhaps we should invite some of the English students to join us, and to inject some of their knowledge into the discussions.'

study group (handwritten marginal note)

We all thought this was a good idea, so George proceeded to organize a meeting. A number of the other research students expressed an interest in the proposal, so it was quite a big group which assembled for the first gathering, in one of the Cavendish labs. However, the meeting was not a success— partly perhaps because there were too many people. George reckoned that the English students were too inhibited to be able to admit openly to, or raise for discussion, things they didn't understand. They tended rather to launch into discourses on topics with which they were familiar, at a level which was in most cases uninformative to us.

'This is a dead loss,' pronounced George, when the four of us got together afterwards to discuss the situation. 'But my guess is that these meetings will soon fizzle out. Suppose we wait until they do that, and then, after a decent interval, we quietly start up a group just for us four nit-wits, to get down to the real question of difficulties and problems.'

So was born the *Nuclear Nit-Wits Club*. We agreed that it should be kept secret from the rest of the Lab, and that we should therefore meet in each other's rooms rather than in the Cavendish. The meetings were usually held in either George's or Charlie's rooms (Charlie Barnes was the other Canadian) in Emmanuel College, these being more convenient than either Godfrey's digs or mine. Thanks to George's drive and our joint enthusiasm for the enterprise, the Nit-Wits Club was a great and continuing success. George had an extremely sharp, analytical mind, and could with admirable perception pin-point basic, as opposed to trivial, problems. He would bring along a recently published paper on some aspect of nuclear physics which he considered interesting and worth thrashing out until we understood it. We would work through the paper step by step, pooling our resources, until we could follow the principles and the arguments, or else identify points requiring further reading. We all did our homework, taking turns to produce problems, and gained increased confidence and satisfaction each time fresh enlightenment dawned. We also discussed specific questions which arose in our own experimental projects as these evolved. George contributed his sharp wit and drive; Charlie his perseverance in getting down to the basics of a problem, and worrying at it until it had been unravelled; Godfrey, with his wartime naval background, was insistent on self-discipline and the ordering and justification of each step. I'm not sure what I contributed—enthusiasm perhaps, and a sense of fun in the whole enterprise, helping to keep things going. Each of us seemed to have a ready sense of humour which enlivened the proceedings. I felt that I received more than I gave to the discussions, but I was always made to feel an equal member.

No one else in the Lab ever found out about our secret society. One more member was added in the autumn of 1947: Allan Cormack, who had just arrived from South Africa. Being another "colonial", under the same educational disadvantages as ourselves, he was considered eligible, and became an active and congenial member of the Nit-Wits. Beneath an apparently care-free manner, he had a keen, original mind, and often contributed pertinent, far-sighted comments.

I look back with great affection and nostalgia to the old Nit-Wits Club, which of course was automatically disbanded when we all left Cambridge: not only was it an invaluable aid to our self-education, but also it engendered in the five of us a warm spirit of camaraderie, and helped to forge bonds of friendship which have lasted all through our careers.

Its members have all distinguished themselves professionally. George Lindsey rose to the position of Chief of the Operational Research Establishment of the Department of National Defence in Canada. Charlie Barnes became a senior Professor in the well-known American University, Cal Tech (California Institute of Technology); his elegant and meticulous experiments, relating to studies of the astrophysical synthesis of the natural elements, were important to the work for which Cal Tech's Professor Willie Fowler won the Nobel Prize

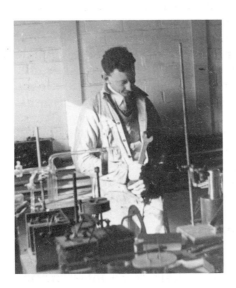

Figure 8.1 Members of the Nit-Wits Club. Above: George Lindsey (left) and Charlie Barnes were also members of the Cambridge University Ice Hockey Team. Below left: Godfrey Stafford in his lab at the Cavendish; right: Allan Cormack.

for Physics in 1984. Godfrey Stafford became Director of the Rutherford Laboratory, Britain's National Centre for high energy and elementary particle physics research, and was later elected Master of St Cross College, Oxford. Allan Cormack—good old Allan, with his infectious laugh and easy-going manner—who for most of his career has been Professor of Physics in Tufts University, Cambridge, Massachusetts, surprised us all, himself not least, by jointly winning the Nobel Prize for Medicine in 1979. The award was in recognition of his brilliant conception of the x-ray based CAT scanner (Computer-Assisted Tomography) which has become a tremendously valuable tool for medical diagnostics. Allan, who worked on this device in his spare time rather than as his major research project, says that one of the aspects of the award which thrilled him most was the receipt of many personal letters of gratitude from people whose lives had been renewed, or in some cases saved, through the use of his scanner.

Apart from Christiane Clavier and myself, there was one other woman research student in the Cavendish amongst the sixty-odd male students. She was June Broomhead, a bright and attractive English girl who began working in the Crystallography Department at about the same time that I started in Nuclear Physics. We soon became good friends, and June's warm and lively companionship during my Cambridge days was a great boon to me. She was what I called a "real" Newnham student—she had been in College as an undergraduate, gaining her Cambridge B A in physics. She was capable and knowledgeable, and had a well-developed critical faculty, which she could apply not only to herself, but also, often with entertaining effect but without malice, to other people in the Lab, and their activities. She was also a good researcher, as I could detect from the respect in which she was held by her colleagues. But she was always modest about her own attributes and achievements. She confirmed for me the view, which I continue to maintain, that a successful woman physicist does not have to be a blue-stocking, or to have an aggressive personality. Far from it, in fact: enthusiasm, perseverence, and an independent spirit, seem to be more important qualities for a woman working in the male-dominated world of physics.

Through June I got to know a number of the crystallographers, some of whom were to become famous. Under the direction of Professor Bragg, the Crystallography Department was a lively group, exploiting the techniques of x-ray diffraction in crystals to work out the atomic structures of highly complex molecules. It was here that Crick and Watson were soon to make the brilliant discovery of the structure of DNA—the material out of which the genes are made, and the key to the process of heredity.

June showed me how, for her experiment, she had to grow a perfect crystal, glue it to the edge of a small piece of glass, and mount it in front of the x-ray machine. She once dropped a good crystal on the floor: it was only a millimetre in size, and she had to crawl about for hours to find it again. The beam from the x-ray source, passing through the crystal, produced a complicated pattern on a photographic plate. With a number of such pictures, taken at different angles, she had to do long, tedious analyses from which the distributions and nature of the atoms in the complex molecule could be inferred. Great perseverance was required for a successful interpretation. But the exhilaration of seeing the solution finally appear was obviously tremendous, as I realized when June came to me one day with shining eyes to show me what she had achieved.

After gaining her PhD, June was recruited by Professor Dorothy Hodgkin at Oxford University, and worked with her for two years on the Vitamin B12 molecule, which had a particularly complicated structure. 'She seems to have a sixth sense,' June commented to me. 'She can perceive all sorts of things in my analyses that are not obvious to me.' Dorothy Hodgkin was indeed a brilliant crystallographer. She had already solved the structure of penicillin; her work on B12 was recognized with the award of a Nobel Prize. I met her once, through June, and was captivated by her gentle, unassuming nature, and her encouraging interest in other people's work—including mine. But I could also detect in her the intellectual power and strength of purpose which made her such an outstanding scientist.

In 1951 June married George Lindsey—George, the vociferously self-avowed misogynist, who used to entertain us during interludes in our Nit-Wits Club discussions with colourfully worded declarations of his astonishment at the time some people wasted chasing after women, instead of getting on with more worthwhile activities. June and George, with their contrasting personalities, but sharing a lively sense of humour, made a good combination and, when they had settled in Canada, produced two outstandingly bright children. June had

Figure 8.2 June Broomhead and George Lindsey (c 1950).

very definite views on the overriding responsibility of a mother for the proper rearing of her family, and gave up her immediate scientific career to that purpose. Later she exploited her teaching talents as a Demonstrator in Physics at Carleton University.

Life for me in Cambridge was not of course devoted entirely to physics. The distractions seemed endless. There were possible talks to attend on all manner of subjects. Tea and evening parties, where carefully hoarded luxuries like cake were produced, provided a full social life. I derived special pleasure from cycling through the gentle Cambridgeshire countryside; the roads in those days were remarkably free of traffic, and when there were three of us we could mostly ride side by side, except occasionally when a passing policeman would call out over his loudspeaker: 'Only two abreast on the road, please.' There was also the idyllic experience of gliding along the Backs in a punt—and the fun of learning how to manipulate the punt-pole. I saw Shakespeare come to life in the romantic setting of a College garden, and realised for the first time that Shakespeare was for seeing and hearing rather than for reading.

Above all there was music: Mozart doubly enchanting against the mellow background of a College quadrangle; the

exquisite singing of the choir boys in King's College Chapel;
world-famous performers in subscription concerts at the Guild-
hall, where I once heard the incomparable Kathleen Ferrier,
her effortless voice a rippling stream of melodic delight. (Alas
she was to die so young.) On another occasion the French
pianist, Cortot, came: a quiet other-wordly man, whose sensitive interpretations of Chopin reminded me, with bitter-sweet
nostalgia, of my mother's playing.

A most memorable event was the special singing by the
Madrigal Society, towards the end of "May Week"—the celebration, marking the completion of the University year, which
seemed, in keeping with the contrariness of many Cambridge
customs, to take place during the first two weeks of June. It
was a balmy spring evening. The singers were gathered in
tethered punts under the arch of King's bridge. The whole of
Cambridge seemed to be there, assembled in punts lining the
sides of the river, or sitting along the banks, down as far as
Clare Bridge. As the singers prepared to start, a complete
hush descended, broken only by the occasional quack of a
startled mallard. Then the traditional songs began, filling the
air with pure, magical sound in the gathering dusk. For their
last song, just as the sun was setting, the performers lit
candles in their punts and then, casting off the ties, they
drifted slowly downstream, the sound of their melody
gradually dying away as the twinkling spots of candlelight
disappeared into the last glow of the evening sky.

Cambridge is a truly unique and beautiful place, I thought.
I'm lucky to be here.

Meantime progress was being made towards the experiment
on the HT1 accelerator. Having insinuated myself into my
present position, I felt a special necessity to show myself in a
good light to Dr Burcham, though I was by no means confident that I could handle all the required operations. However,
Burcham was very understanding, and provided gentle guidance and ready help. Some of the equipment which he had
used in his pre-war experiment was still available, the electromagnet in particular. He pointed it out to me, in a corner of
the HT1 target room. 'It's just a laboratory magnet,' he said,
casually. I scrutinized the large, robust, venerable-looking
object with some awe, wondering what important functions it
might previously have fulfilled, during the Rutherford era. It
stood about three feet high, mounted on a wheeled base, with
two coils of many turns of heavily insulated wire wound round

a massive U-shaped iron yoke. At the top the yoke carried two flat-faced ten inch square iron pole-pieces, separated by a vertical $1\frac{1}{2}$ inch gap, into which a rectangular copper vacuum box was fitted.

A new target chamber was required, and this was being made in the Workshop to Burcham's design. Alex Baxter was working on the ionization chamber and the associated electronics. One of my responsibilities was to make a very thin mica window for the exit port of the vacuum box. Using a sharp razorblade, I learned painfully the art of lifting off wafer-thin layers from a piece of mica, and measuring their thickness and uniformity with a radioactive alpha-particle source. These almost invisible little pieces of mica were very fragile, and could have blown away in a draught. I hardly dared to breathe when I came to the point of sealing one into the vacuum box. Then I pumped the box out very gently, to check that the window would withstand atmospheric pressure: the vacuum held, and I heaved a sigh of relief.

In due course all the components of the apparatus were ready for assembly in the target room. It was at this stage that I began to appreciate the tremendous value of Willie Birtwhistle, the Technical Research Assistant in charge of the HT1 set and the target area. He was a quiet, chubby, strongly-built north-countryman with a delicious Yorkshire accent, generally serious, but with a ready sense of humour, and a kindly, lovable personality. As he told me later, he was trained in electrical engineering and was brought to the Cavendish by Cockcroft in 1930, helping to build the Cockcroft–Walton accelerator which made history in 1932. The assembly and commissioning of the HT1 set had been largely Birtwhistle's responsibility, as had been its operation and maintenance during the war years and after. He knew the machine in every detail, and was clearly dedicated to his job. Above all, from my point of view, his helpfulness and patience with the students were boundless. With quiet efficiency he helped me to assemble my equipment, mounting it all on the big electromagnet so that it could be wheeled into position in front of the proton beam pipe when the time came to run the experiment.

Then, at last, the great moment arrived. That beautiful, futuristic-looking machine, standing in its cathedral-sized hall, was going to operate just for me. Birtwhistle moved the heavy equipment up to the accelerator beam pipe, coupled the target chamber to it and switched on the vacuum pumps. For what seemed an interminable time he watched the pressure readings creep down. 'That's good enough,' he finally decided. 'We can open up to the accelerator now.' He disappeared into the big HT hall. Watching him from the doorway I noticed how agile he was for a heavy man, climbing quickly up the small ladder onto the target room roof to make adjustments to the machine and the ion source.

Returning to the target room, Birtwhistle turned on the power supply for the analysing magnet which would bend the proton beam, coming down from the accelerator above, into a horizontal direction along the beam line to the target chamber. Then he slowly turned up the machine voltage, watching the meter which would register a proton beam current to the insulated target holder. The needle of the meter started to flicker.

Suddenly there was a tremendous bang, like a great clap of thunder, from the HT hall. I nearly jumped out of my skin. 'It's all right,' said Birtwhistle calmly, as he wound down the voltage control quickly. 'We often get spark breakdowns like that at the higher voltages. The pressure's gone bad, so we must have sprung a leak in the accelerator tube when the spark went off. I'll just go in and fix it.' He disappeared into the HT hall. I watched him first discharge the terminal with an earth lead fastened to the top end of a long insulated pole. 'There's always a bit of residual charge on the terminal,' he explained as he drew a fat spark from it. Then he was up a ladder, thumbing with firm stroking movements the soft black plasticine which surrounded one of the joints in the accelerator tube. I noticed that there were many joints where this special plasticine had been applied. 'How did you know which joint to go to?' I asked. 'Oh, you get to know,' he replied nonchalantly. 'A big spark like that causes a sudden pressure surge and the plasticine is lifted off at a weak spot. There's never been time to repair all the faulty joints properly.'

Back in the target room we found that the pressure had come down again and soon Birtwhistle had the current meter registering a proton beam on the target. 'You've got thirty microamps on your target now,' he said. 'That should be enough to give you some results.'

So it was over to me. My hands trembling slightly with excitement, I slowly turned up the power supply to my electromagnet, looking for counts from the ionization chamber. I expected them to appear when the magnetic field reached the right value for bending first, scattered protons, and then alpha particles, into the direction for entry into the chamber.

Now there should be some proton counts. Nothing happened. I increased the field a bit more. No counts. I tapped the body of the ionization chamber gently: the amplifying system was so sensitive that this mechanical shock was enough to make it register a splurge of counts. The ionization chamber must therefore be working properly. I increased the magnetic field to its maximum value. Still no counts. Sweating with anxiety I turned the magnetic current down again very slowly. There was nothing to be seen in the way of particle counts. I looked at Birtwhistle, who was watching my procedures with solicitude, and thought of the monster machine next door just waiting on me. What on earth was I to do? My mind seized up in panic.

Then Dr Burcham came into the target room, snatching a few minutes between his lectures and other duties to see how I was getting on. Miserably I told him of my failure to see any counts, and described all the procedures I had followed. He pondered for a minute or two and then smiled his slow, imperturbable smile.

'Perhaps you have connected the power supply to the electromagnet the wrong way round,' he suggested.

'Oh horrors!' I muttered to myself. 'I should have thought of that.' If the current in the magnet coils had been flowing in the wrong direction, the magnetic field would have been bending all the particles away from rather than towards the ionization chamber. This had to be the explanation. I looked at Burcham and nodded slowly in sheepish acknowledgement. He, still smiling in a slightly embarrassed way, glanced at his watch and went off.

Birtwhistle and I had soon reversed the leads of the power supply and got the experiment started up again. Once more I turned up the magnet current, and soon the proton counts came pouring in; at a slightly higher current I could see some bigger pulses from alpha particles. We were in business at last.

'Joan Freeman, you are a blithering idiot.' I exclaimed aloud, thinking how foolish I had made myself appear in Dr Burcham's eyes, to say nothing of wasting precious accelerator time.

'Aw, I wouldn't worry about it,' said Birtwhistle, his soft Yorkshire accent somehow emphasising the comfort of his advice. 'People are making much more stupid mistakes than that every day.'

I did indeed soon forget about it in the excitement of proceeding with my first experiment on the HT1 set. At last I was doing what I had come to Cambridge for. Moreover I had a real Supervisor whose effectiveness had already been well proved; I had a great ally in the inimitable Birtwhistle; and I had the back-up of the technically highly competent Baxter. Soon I was measuring the ranges of the alpha particles from the reaction of protons on fluorine, and in due course I had succeeded in confirming all the results which Burcham and Devons had previously obtained. I was then ready to break new ground, studying reactions which had never been observed before. Life was now indeed full and sweet.

Chapter 9

On Target

Towards the end of 1947 an appointment was made to the vacant Jacksonian Chair in Experimental Physics. The occupant was to be Professor O R Frisch, a name well-known to nuclear physicists. He was an Austrian, of Jewish origin, who, before the war, had worked in Germany and then Denmark, before being driven in 1939 to take refuge in England. His most famous achievement had been the interpretation in collaboration with his aunt Lise Meitner, also a remarkable physicist, of the momentous discovery published early in 1939 by two German chemists, O Hahn and F Strassmann. They reported that, following the bombardment of uranium by neutrons, much lighter elements, like barium, were found. The conclusion reached by Frisch and Meitner was that the uranium nucleus was being split into two more or less equal parts, with the release of a large amount of energy. Frisch called the process nuclear fission, and, with his attractive sense of humour, referred to the products as "fission chips". Later, at Birmingham University, Frisch, with the brilliant German-born theoretical physicist Rudolf Peierls, worked out that an atomic bomb was possible. Between 1943 and 1945 he was at Los Alamos, where the first atomic bomb was made, and from 1945 to 1947 was Head of the Nuclear Physics Division of the newly set up Atomic Energy Research Establishment at Harwell.

Frisch appeared in the Cavendish early in 1948. He looked very much an intellectual, with broad brow and a thoughtful, faraway expression. He wandered about vaguely, without communicating, as if somewhat lost. I thought at first that he must be shy, but concluded later that it was more a case of indifference unless his interest was aroused. He was certainly a very unconventional character. Shortly after his arrival, as I was cycling towards the Cavendish over the busy Silver Street Bridge after a sudden rain storm, I saw him stooped over a puddle, his cycle lying in the road beside him. Halting, I asked if I could help. 'Oh, no,' he answered brightly, and without embarrassment, 'I'm just rescuing a ladybird.' Later I noticed him sitting on a seat which faced the Silver Street traffic lights, near the Guest House where he was staying. He was fully absorbed in a paper-back novel with the green cover characteristic of detective stories.

Figure 9.1 Professor O R Frisch (c 1954). (Courtesy of Mrs Ulla Frisch)

Frisch usually attended the colloquia given regularly by staff members or students, when he would stare disconcertingly round the room, or make sketches of the speaker—he was extremely skilled at drawing caricatures—or just go to sleep. Yet at the end of the talk he would produce some very pertinent and telling comment. He was imbued with a love of physics, and his intellectual ability and knowledge commanded our unqualified respect. However, it soon became clear that we were not to find in him a dynamic leader. His completely unselfconscious, almost childlike, manner was endearing, but he was irresponsible about carrying out jobs that he did not want to do.

It was not long before I encountered an extraordinary example of this characteristic. Having obtained some original results on my short-range alpha-particle experiments, I was encouraged by Burcham to write a short paper, in collaboration with Baxter. The preparation of this, my first publication in the scientific literature, was an exciting task. When it was complete, I had, in accordance with a long-standing departmental rule, to have the manuscript read and approved by the Professor before submission to the Journal. I approached Frisch several times about it, but his response was always that he was too busy to look at it just then. I knew that it was no good just leaving it on his desk, which he seldom seemed to occupy. What was I to do? I decided to lurk in the alcove on the second-floor corridor, from which I had a view of his office door, and wait until he appeared. In due course I saw him amble along the corridor and go into his office. I knocked boldly on the door, went in, and placed the manuscript on the desk in front of him, asking him to look at it right away. Grunting, he stared at it for a minute or two. Then suddenly he opened a desk drawer, drew out a pair of trousers, glanced at his watch, and said 'I must take these to the cleaners.' With that he was gone. My friends hardly believed my tale until, some time later, Charlie Barnes reported that exactly the same thing happened to him when he tried to tackle Frisch on some problem. Evidently the trousers were kept permanently in the drawer for such emergencies.

Charlie was in a worse predicament than I was. Having initially, like me, been without a Supervisor, he was allocated to Frisch on the latter's arrival. Charlie went on struggling for a while with the project that he had been tackling, and then decided to consult Frisch. After listening to Charlie for a minute or two, Frisch suddenly asked 'Who is your Supervisor?' 'Well, you, Sir,' Charlie replied diffidently. 'Oh, am I?' responded Frisch, quite unabashed.

Frisch's main interest in the Lab seemed to be that of designing original scientific devices. As well as offering useful ideas to other people, he made some very ingenious instruments himself. One of these was a mechanical kicksorter: a forerunner to the electronic pulse-height analyser soon to be widely used for recording the spectrum (numbers versus pulse height) of electrical pulses produced by nuclear reaction products in a detector. It was an inclined board with thirty parallel grooves. Each electrical pulse was made to actuate a billiard cue tip, which gave to a little steel ball a kick proportional to the size of the pulse. The ball described a parabolic path across the top of the board and then dropped into the groove which corresponded to its initial velocity. As balls, fed from a hopper above the board, were fired by successive pulses, a spectrum was built up in the grooves to represent the pulse-height distribution. It was a pretty device which worked remarkably well, though it could only cope with counting rates of six a second or less. I used it successfully in one of my experiments.

Much later Frisch designed and built a sophisticated machine he called the SWEEPNIK, which is now used all over the world by high-energy physicists for analysing elementary particle tracks, and which won a British industrial award for its inventiveness.

Frisch was also musically gifted. At our annual Cavendish dinners, which he clearly enjoyed, he would play classical music for us on the piano with skill and great sensitivity. And on Friday evenings he was sometimes seen waiting at Cambridge Station for the London train, clutching a violin case; however his motives were not entirely musical: in the spring of 1951 we learned that he was marrying a Viennese artist who lived in London.

One Saturday morning Frisch unexpectedly brought his famous aunt, Lise Meitner, into the Lab, to show her around. I found her quite fascinating: a tiny person, with bright, penetrating eyes and a dynamic personality, taking an intense and critical interest in everything she saw. Since it was a Saturday there were not many people about. 'Where are all the scientists?' she demanded of Frisch. 'Well, it's Saturday,' he said, by way of explanation. She snorted her disapproval, implying that dedicated researchers should be wanting to work whatever day of the week it was. She took a particular interest in me, and wanted to know how I had come to take up physics, as well as following closely my demonstration of what I was doing. I could well believe that she had been a driving force and inspiration to everyone who had worked with her.

During 1948 my short-range alpha-particle experiments progressed at an exciting rate, particularly after two major improvements were made to the apparatus. One of these was the introduction of a photoelectric multiplier which Baxter had managed to acquire from EMI. This was set up in place of our ionization chamber, with a fluorescent screen in front of it, to detect the alpha particles; a pre-amplifier built by Baxter in one of his tobacco tins was mounted at the base of the multiplier.

Figure 9.2 The magnetic spectrometer designed by Dr Burcham. The accelerator beam pipe leading to the target chamber can be seen top back. In front is the cylindrical housing of the photoelectric multiplier with Baxter's tobacco-box containing the pre-amplifier at its base. The apparatus stands about three feet high.

The other, very important, innovation in our apparatus was a magnetic spectrometer, devised by Burcham. It was designed not merely to remove unwanted protons scattered from the target but to act as a magnetic lens, bringing to a focus at the fluorescent screen alpha particles emitted from the target at a particular energy. The same laboratory magnet as before was used, with special pole-pieces and a length of rectangular copper waveguide for the vacuum box. I did not realise at the time what a pioneering instrument this was for the study of charged particles from nuclear reactions. Our simple instrument was in fact one of the first in a long line of magnetic spectrometers, represented nowadays by very large, sophisticated, and expensive pieces of equipment. I later learned that at two nuclear physics laboratories in the USA (MIT and Cal

Tech) magnetic spectrometers, of a somewhat different design, were starting to be used for similar studies at about the same time.

The method proved to be very effective for studying proton-induced reactions with a variety of targets, and, with Burcham taking an increasing interest in the work, we obtained significant new results for a number of light nuclei.

1948 was altogether a full and exciting year for me. One highlight was the emotional reunion with my mother, when she arrived on leave of absence from her Pymble school. She found a temporary job teaching mathematics at a school in Hitchin, and shared most of her holiday periods with me. For one of our excursions I chose a cycling tour of Cornwall. My mother had not cycled for about fifty years, but, game for anything, she pedalled enthusiastically, with a rucksack on her back, up and down the Cornish hills, relishing with me the charm of this unique corner of Britain.

Later we discovered the Continent: the fascination of Paris; the special delights of Switzerland with its tiny, fairy-tale

Figure 9.3 Ada Freeman on an excursion with the author in Switzerland.

villages and the breath-taking splendour of the Alps; Norway, with its remote grandeur; and Rome, soaked in history. Italy had only just been opened to tourists when we went, and Rome, uncrowded and unspoilt, was particularly memorable, with its dramatic revelations of the Renaissance and Roman periods; standing amidst the ruins of the old Roman Forum, oblivious of the mid-day August heat, I visualized those ancient times and saw, in my stimulated imagination, characters like Caesar and Cicero from my dry old Latin textbooks suddenly coming to life. I felt that I could spend a lifetime exploring the wonders of Europe.

It was also in 1948 that I began to get to know John Jelley. He was one of Burcham's students: a short, sturdy young man, with a ruddy complexion and fair wavy hair. He worked with a Canadian called Eric Paul, who was very tall and thin, with a long, pale, hollow-cheeked face and wispy brown hair. They looked an ill-assorted pair as they strode about together with serious and purposeful expressions, neither of them particularly communicative. They were already engaged in a complicated-looking experiment on the HT1 set when I first arrived

Figure 9.4 John Jelley (right) with Bill Burcham at the Cavendish in 1948.

at the Cavendish and I had felt shy of approaching them. But then, one day, while I was in the target room setting up for a run, John came in on his own and, perching on a stool, started to chat to me.

He told me that, although his parents were both artists—his father already sixty when he was born—and he had had a restrictive upbringing as an only child, he developed a spontaneous interest in physics at an early age. He enjoyed designing and constructing models and mechanical devices; in his early teens he made a crystal radio receiver, and later, a valve set with which he could listen to radio transmissions from round the world—including one from Australia with a kookaburra's laugh as the call sign, he told me. Then, while television was still very new, he built himself a receiver which successfully recorded the 30-line images.

Astronomy was another abiding interest for John. He built an eight-inch mirror telescope in his back garden and used it to observe the major planets and their satellites. He said that he had considered studying astronomy after graduating in physics from Birmingham University, just before the war. But he had been swept up, first into radar research, and then into atomic energy, being sent in 1944 to Cockcroft's Laboratory in Canada. On his return two years later he registered as a Ph D student at the Cavendish, choosing to go into nuclear physics. 'But sometimes I wonder why I am cooped up in a lab,' he suddenly said, looking round the cluttered room as if he felt trapped. 'I love the great outdoors, and want most to be in wild, unpopulated places.'

John seemed an unusual mixture: on the one hand he gave the impression of being a loner, somewhat naive and socially undeveloped. But at the same time there was about him an engaging openness, generosity, and tolerance. After his narrow existence as a boy, he had clearly experienced a dramatic emancipation when he went to Canada. He talked with tremendous enthusiasm of the varied and interesting characters he had met there, and of skiing weekends in the Laurentian hills, excursions to the Rocky Mountains, and visits to the United States, particularly New York. He loved the Canadian climate and the free-and-easy way of life. An intriguing young man, I thought. It would be interesting to get to know him better. He was in fact to give me ample opportunity for this.

By an interesting coincidence, during 1948, one of my few known relatives arrived in the Cavendish. This was my first cousin, Henry Dyer, son of my mother's younger sister, Irene. In her early twenties Irene had decided to leave her teaching post in Perth, to visit England, and to see the world. But the first World War cut short her voyage: she got only as far as South Africa. She took a teaching job in Johannesburg, there married an English engineer, Wilfred Dyer, and had two sons. Her life in some senses paralleled my mother's: her husband lost his job in the Depression, and Irene made ends meet by taking on and successfully running a hotel, as well as producing and selling some fine landscape paintings. Neither she nor my mother were good correspondents, so it was from out of the blue that her younger son, Henry, having won a Scholarship from the University of Natal, appeared at the Cavendish to do a Ph D in Crystallography. Tall, red-headed, and somewhat shy, his two passions were physics and rugby, at both of which he worked hard and effectively. I enjoyed the novelty of meeting a real relative, and wondered if there was a tendency to physics somewhere in the genes. After returning to South Africa with his Ph D, Henry joined the Industrial Diamond Division of de Beers, was a member of the team of four who in 1959 succeeded in producing synthetic diamonds, and ultimately became Managing Director.

By the end of 1948 I had begun to approach the task of writing my thesis, and to consider the problem of my future. I had gained the agreement of the CSIR for me to continue at the Cavendish for my third year as a research student; they had allowed me a small additional grant which, combined with a Newnham College Studentship, was sufficient to keep me going financially until August 1949. But then what? I felt a strong sense of obligation to return to the CSIR, but no immediate desire to do so. After the big initial struggle to get into nuclear physics at the Cavendish, I now felt immersed in it, and driven by the fascination and promise of the work. I wanted to carry on with it, for a while at least. In Australia there appeared to be very little opportunity for nuclear studies. If I returned to the Radiophysics Lab I would probably go into radioastronomy. This, under Pawsey's leadership, was undoubtedly an exciting field, which I knew was progressing at a remarkable rate. Ron Bracewell, planning to return at the end of his third year, was in a good position to drop straight into this slot, having spent his time at the Cavendish in radiowave research. But for me, with quite different experiences in Cambridge, it would be a much more difficult transition. On the other hand Burcham, by now very enthusiastic about the way in which the short-range alpha-particle studies were developing, was urging me to continue with the work after my Ph D, saying that a temporary post as a Cavendish Research Assistant could be arranged for me.

This prospect, and the lure of unexplored wonders in Britain and Europe, were a very strong inducement to me to stay. I wrote to Dr White and Dr Bowen putting my feelings to them. They were very understanding, and agreed that I could continue at the Cavendish for a year or two on leave of absence. So, setting aside my residual sense of guilt, I proceeded to throw myself into the experimental work and thesis writing.

During 1949 I changed my digs to the home of Pat and Harold Whitehouse. Pat, a Newnham contemporary of June Broomhead, and Harold, a Lecturer in the Botany Department, had recently married and had acquired a large, comfortable, three-storied Victorian house on the Barton Road, providing accommodation for three or four Newnham students. I revelled in the easy-going atmosphere of the Whitehouse home and appreciated the warm friendship which Pat and Harold extended to me. Pat was to have a remarkable career. She gained a medical degree, as well as a Ph D in Natural Sciences, but photography was her overriding interest. She specialised in stereoscopic colour photography, particularly of birds, insects and flowers in close-up. Drawing on her considerable scientific skills, she designed and made her own unique stereo cameras, and produced exquisite and highly original audio-visual shows which have been in demand all over America and Europe as well as Britain. In 1972 she was awarded the prestigious Hood Medal of the Royal Photographic Society. Yet she was a modest and gentle person, of whom I became very fond.

In 1951 Pat had a daughter, Jane, to whom I was invited to be godmother. This relationship has been a source of great pleasure and satisfaction to me. Not having had children of my own, or even the possibility of nieces or nephews, Jane has been the nearest equivalent. I kept in close touch with her as she progressed from childhood to womanhood, gained a University degree, took up teaching, and married Phil Cooper, an architectural engineer. They now have three children, and the family live in an interesting old house in Cambridge where I frequently have the pleasure of staying.

Writing my thesis, to say nothing of preparing some sixty

figures, was a big effort, and, in that era before the photo-copier, decidedly tedious. But finally three bound copies of the thesis were ready for submission. Meantime my appointment as a temporary Research Assistant was approved. Burcham, giving me this news, explained rather apologetically that the assent of Mr Shire, who was still in administrative charge of the Department, had been required, and he had insisted on the proviso that I undertake some specific job in addition to the research on HT1. 'So I suggested that you could if necessary design and build a double-focussing magnetic spec-trometer, like the one they have at Cal Tech,' Burcham said. I was taken aback. This would be a big engineering project which I did not feel qualified to undertake. 'I'm sure you could do it,' said Burcham, 'but I wouldn't worry about it. I don't suppose it will eventuate, because our present spectrometer is quite adequate for our purposes.' So I dismissed my concern, concentrating on the experimental work and on a novel diversion which Burcham proposed.

While working with me on a paper about the proton bombardment of an isotope of boron, Burcham remarked that it would be interesting to study the short-range alpha-particle emission following the bombardment of the same target with neutrons. He told me that this might be possible at the Atomic Energy Research Establishment, where a recently commissioned pile (the name then given to a nuclear reactor) was reported to be producing neutron beams for experimental purposes. 'If you like, I'll write to Cockcroft and ask him if it could be arranged for you and Birtwhistle to take your apparatus over to AERE and do the experiment there,' he said.

I leaped at the idea, and Cockcroft readily agreed. Thus, arriving with Birtwhistle in a van into which he had loaded all our equipment, I had my first sight of AERE—popularly known as Harwell, to the indignation of the inhabitants of the nearest village, which had borne this name for centuries.

The Establishment was located at the site of an important wartime RAF aerodrome, in attractive agricultural country close to the Berkshire Downs, about twelve miles south of Oxford. It looked to be a huge development, with many scattered buildings, some evidently ex-RAF and some in the course of construction, surrounded by a formidable barbed-wire topped fence which proclaimed the tight security imposed on the Establishment. A police guard at the big iron gate directed us to the Lodge where the checking of our credentials and the issuing of passes took place. We were then conducted past three great aircraft hangars to a fourth, in which BEPO (British Experimental Pile) was housed. It was indeed an enormous cubic pile, with large pieces of equipment set up on top and all round it. I realised that its great size was dictated by the requirement of heavy shielding from the neutrons and gamma radiation produced in the reactor core.

Feeling dwarfed and over-awed, we were taken round to one side of it by our deputed guide, a rather taciturn engineer, who gave us the instructions and help we needed, but seemed quite uninterested in us and our project. He led us to a gaping hole a few inches square in the side of the reactor, about three feet from the floor. This, he said, was the opening of a channel from the reactor core, and through it an invisible beam of thermal-energy neutrons was being emitted, to be absorbed in a large concrete block placed some ten feet back from the pile face; the hole could be plugged temporarily with a shielding block while we set up our target and detection equipment in front of it. He cautioned us to avoid standing in the path of the beam when the plug was out. I suspect that our guide may have had a laugh at our expense, because, in walking towards our allocated beam hole we had to pass another apparent hole about a foot from the ground. He said that we must be sure to step over the beam coming out of that. So, repeatedly for the three days while we were working there, Birtwhistle and I had to do a high step over this invisible, and, I suspect, hypothetical beam; I looked for, but did not manage to detect, suppressed laughter from the various groups of men working at or near the pile face.

However, having got our equipment set up, we were delighted to see our detector recording short-range alpha particles as a result of the neutrons hitting our target. Having been used to needing a vacuum pipe to convey the proton beam from our HT1 set, it seemed strange to have these little neutral particles streaming unabsorbed through the air into our target chamber. The experiment was remarkably successful and Burcham was duly gratified with the results. He arranged a second visit to AERE for us in the following year, to an accelerator called a Van de Graaff, which had just been commissioned in another big hangar. This time we received a warm welcome, the small accelerator group being genuinely pleased to have a university user of their new machine, and I gained a favourable impression of the liveliness of the Nuclear Physics Division of which this group was a part.

While there, I was invited to give a talk to the Division on my short-range alpha-particle work. The audience was comfortingly receptive and tolerant, without the competitive and over-critical attitude which research students at the Cavendish tended to adopt. One member of the audience in particular, an Italian of great charm, made some perceptive and encouraging comments, which I appreciated. I learned that he was the leader of the Cosmic Ray Group, which John Jelley had recently joined. His name was Dr Bruno Pontecorvo. A brilliant physicist, his popularity with the Group, which he had initiated and inspired, knew no bounds. They were devastated and completely mystified when, in September 1950, he suddenly disappeared from the Western World. They refused to believe that he had been a spy, John told me.

The dreaded Ph D oral exam took place in December 1949. The internal examiner was, by tradition, the student's Supervisor, in my case Burcham, who was typically kind and gentle on the occasion. But my external examiner was a formidable character: Dr Egon Bretscher, Head of the Nuclear Physics Division at AERE. I had already heard of him as a very able scientist, with a tough, uncompromising personality. Swiss by birth and early training, he had worked in the Cavendish under Rutherford and became a Lecturer in Physics at Cambridge. During the early part of the war he had been in charge of the team studying neutron reactions using the HT1 set; later he went to the USA to work on atomic bomb development at Los Alamos. He became well known for his work on the element plutonium. After the war, he took a senior post at Harwell.

Just before I was to present myself for the oral, Burcham came to me apologetically with the warning that Bretscher had arrived complaining of a migraine headache, was insisting on keeping his overcoat on, and was very grumpy. It was just as well that I had had this warning. His appearance was intimidating as I entered the room. He was very tall and lean, with a long thin face, prominent nose, a high forehead practically bald on top, and penetrating eyes with which he fixed me disconcertingly. But he greeted me with European style courtesy, and, when he began to speak, with a strong Swiss accent, his voice was deceptively quiet and gentle. He had obviously read my thesis carefully, and his questions on it were sharp and penetrating, probing for things I wasn't sure of rather than asking about topics I had hoped to expand on. Then came tough general questions, culminating in his asking me what the electric quadrupole moment of the deuteron was. I hadn't the faintest idea and said as much. But he wasn't going to let me get away with that. 'You must work it out,' he said. It was like having a tooth pulled. Painfully, step by step, he extracted out of me a calculation from first principles; and, having reduced me to a state of mental paralysis, he looked positively triumphant. What a man, I thought, with a sense of mingled fear and respect, little imagining how much I was to see of him in the future.

Fortunately for my peace of mind, I was not left in suspense for long. After their deliberations, Burcham sought me out and told me unofficially that they had both been satisfied with my performance, as well as my thesis, and were recommending me for the Ph D degree. So the pipe-dream of my Sydney University days had come true, to the enormous satisfaction of my mother as well as myself. I was even able to enjoy the privilege of having my degree conferred in a traditional ceremony in the Senate House, wearing the handsome scarlet hood which Denys Wilkinson had lent me. I suppose I must have been one of the first few women in Cambridge formally to take this degree. It was a thrill to be able to write the title "Dr" in front of my name, though I soon learned to use it with caution, particularly after the proprietor of a guest house where I was checking in started telling me about his hernia. One of my colleagues told me that he once overheard his landlady saying to a friend of hers: 'He's the kind of doctor that doesn't do you any good.'

Some time later, while I was working peacefully in my lab, my serenity was suddenly disturbed by the entry of Mr Shire. 'How is the double-focussing magnetic spectrometer coming along?' he demanded. I hadn't given it serious thought since Burcham had implied that it probably wouldn't eventuate, even though it had been mentioned in connection with my appointment as a Research Assistant. I stammered something to the effect that I hadn't done anything about it. 'Well you'd better get on with it,' Shire said snappily, and walked out. With a feeling of despair I found Burcham and told him what had happened. But he clearly didn't want to get involved in the matter. 'Well, I suppose you'll have to go ahead with it,' he said. 'I'm sure you'll be able to cope, with the help of the Engineering Department.'

Looking back, it seems to me a miracle that I did somehow manage to cope with the production of this large 3-ton precision instrument, in spite of having had virtually no real engineering or design experience. The staff in the Engineering Department were in fact tremendously kind and helpful. They agreed to manufacture and assemble all the components for me, and made vital suggestions when I brought my design drawings to them. The spectrometer was to be modelled on one recently developed at Cal Tech in Pasadena, and from them I obtained some helpful information, particularly from Professor W A Fowler, with whom I had already been corresponding about experiments similar to ours that they were doing. But the spectrometer design presented me with many difficulties and I went through some agonizing times with it.

The project also had its amusing moments. One of my problems was that of working out the shaping of the pole-piece edges for the electromagnet. In those pre-computer days this was not easy. Someone suggested that I might be able to make a model using a device known as an electrolytic tank. There was one, they said, in a laboratory on the Madingley Road, where a group was starting to design a linear accelerator for the Cavendish; so I arranged to go and see their tank one afternoon. As I was setting out, just after lunch, there was a sudden heavy rainstorm, so I waited until it was over and then cycled to the lab, which looked like an old factory building. No one responded to my knock at the door, so I opened it and found myself in a large room where all the work was apparently going on. There were about half a dozen men there—in their underpants, staring at me in consternation. Evidently, cycling back after their lunchtime excursion, they had been caught in the rainstorm, and had taken their sodden trousers off and hung them up to dry near a radiator. One of them was D W Fry, there as a consultant from Harwell. I had literally caught with his pants down a future Director of the UK Atomic Energy Authority Establishment at Winfrith. I retreated hastily, but after donning their wet trousers they invited me in again affably enough. A man called Tom Gubbins introduced me to the electrolytic tank, but I realised that it was not accurate enough for my purposes. In the end I worked out a graphical method of solving the problem, and, to my ultimate relief, the design I arrived at proved satisfactory.

One of my big problems was the choice of insulated copper wire for winding the coils of the electromagnet. The insulation would have to withstand the heat generated in the coils, without electrical breakdown. I made an excursion to a prominent wire-manufacturing firm in London's East End, discovering with amazement this extensive industrialized aspect of the Big City. The firm itself, which had written to me on impressive-looking notepaper, also surprised me with its primitive appearance—one large over-crowded workshop with a tiny elevated office accessed by rickety wooden steps. The workmen stared at me as if they had never seen a woman before, but the manager was helpful. With trepidation I ordered a large quantity of what seemed to be a suitable type of wire. It proved to be the weakest aspect of my spectrometer design: after the instrument had been run successfully for some time, I heard that there were breakdown problems. However, I feel that the outcome could have been a lot worse. I was very much a sealing-wax-and-string type physicist, and certainly no engineer.

Early in 1950 I met the famous Dr (by then Sir John) Cockcroft. He had come to the Cavendish on a recruitment drive for Harwell. It was remarkable to me that the Director of a large establishment should come himself for this mission, rather than sending an envoy. But the personal approach, as I later learned, was very much his way. He invited all the recently graduated, or soon to graduate, students of the Nuclear Physics Department to come and talk to him individually, and chose for the interviews the informal atmosphere of the alcove on the second floor of the Austin Wing. I was surprised to find him a very ordinary-looking man with a straightforward, unpretentious manner. He wore a fixed expression, and an enigmatic half-smile which betrayed nothing of his thoughts. But there was about him an air of purpose and quiet determination.

With an exceptional economy of words he told me that Harwell had been set up to investigate atomic energy in all its aspects—physical, chemical, biological, metallurgical, engineering—and this included fundamental research, which was essential in laying foundations for the future. Advanced facilities were being provided, and for nuclear physics there were four particle accelerators. University Departments, which were not individually able to afford such machines, were being encouraged to come and make use of the Harwell ones and to participate in joint research projects with Harwell staff. The research groups, populated by first class scientists, many of whom had had experience in Canada and America during the war, operated on University-like lines, with freedom to pursue good ideas. If I joined one of these groups I would have ample opportunity to carry out some front-line research, Cockcroft said.

His factual statements and persuasive optimism made his proposition sound attractive, but I explained my problem: a sense of moral obligation to return to Australia. 'I'll think about it,' I said, somewhat ungraciously, to one of the most important men in British physics.

During that same year I had the great pleasure of seeing Joe Pawsey in Cambridge. His primary objective at the Cavendish was to talk to Martin Ryle and other members of Ratcliffe's group working in radioastronomy. After that he came to see what I was doing. It was wonderful to see the welcoming, boyish grin again. He took a close interest in my work, characteristically plying me with questions which probed the basic principles and significance of the research, even though this was not his field. Then he told me about his own progress and the latest exciting discoveries in radioastronomy by his group at the Radiophysics Lab: their results were comparable with, and in some cases ahead of, those of Ryle's rival group.

'Now let's go out and enjoy Cambridge,' Pawsey suddenly said. We went punting on the Cam, Joe enthusiastically demonstrating his prowess with the pole. Later I took him to the Arts Theatre, to see a performance of *Swan Lake*. He had never seen a ballet before, but professed himself interested in having a new experience. Sitting in the stalls with his long head and neck sticking up conspicuously, he embarrassed me somewhat by asking occasional questions in an audible whisper about what was supposed to be going on. His summing up at the end was that he enjoyed it but thought that the man didn't seem to have much of a part, being expected mainly to act as a prop for the woman.

While I was with Joe, I took the opportunity of confiding to him my dilemma as far as my future was concerned. On the one hand were my feelings of affection for and loyalty to my native country, and a sense of obligation to return to the CSIR at the termination of my Research Assistantship. On the other, I wanted to pursue nuclear physics, and Harwell seemed an obvious place to go for this. I also confessed to my growing interest in John Jelley, who was about to join Harwell; although matrimony was not being immediately considered, I didn't want to cut short our present companionship by going back to Australia.

Pawsey was very understanding, and said emphatically that I should follow the path that would make me happiest. He did not think that CSIR would worry about the question of obligation, and, very generously, he argued that exchanges between Australia and England were a good thing: Australia was benefitting from the contributions of successful English scientists who had emigrated there, and at the same time good Australians working in Britain served as ambassadors to strengthen Commonwealth ties. What an encouraging and reassuring person he was, I reflected gratefully.

I saw Pawsey once more, in 1954. He had come to England chiefly to be initiated as a Fellow of the Royal Society—a richly merited recognition of his outstanding achievements in radioastronomy, and an honour only too rarely extended to scientists in Australia. The occasion gave him simple pleasure, and he telephoned me from London inviting me to have dinner with him and go to a theatre, by way of celebration. We had a very happy evening, and it was clear that he no longer thought of me as "just one of the boys". That was, alas, the last time I saw Joe. Tragically, in 1962, while still at the height of his career, he died as a result of a brain tumour. I mourned deeply the loss of a friend who was such a gentle, lovable person, and at the same time the most brilliant leader of a research team that I had known.

As 1950 progreesed I felt that it was time for me to move on. My contemporaries, the great wave of nuclear physics students who had started in 1946, had left or were on the point of departure. Baxter had got his Ph D and had set up a vacuum techniques firm in Cambridge; John Jelley had gone to Harwell; Devons had moved to Imperial College; and Burcham was soon to take up a Chair of Physics in Birmingham University. I was fortunate to have been a part of the productive post-war peak in nuclear physics research at the Cavendish. It was soon to decline. Denys Wilkinson, who was making significant contributions in the field, did indeed stay on for a while, anticipating new developments with the linear accelerator which was being constructed on the Madingley Road site. But after Bragg retired, in 1953, his successor, Professor Mott, not interested in nuclear physics, finally decided to close down the project, and Wilkinson moved soon afterwards to Oxford. Meanwhile A P French, a bright young staff member whose friendship as well as his general wisdom and research ability I had come to value highly, left for

America, there to carve out for himself a distinguished career in advancing the standards of physics education. Thus the Nuclear Physics Department was steadily depleted and the fame of the Cavendish was to be furthered in other fields of research.

Encouraged by what Pawsey had said to me, I decided to apply for a job at Harwell, and was offered a post as a Senior Scientific Officer in the Nuclear Physics Division, under Dr Bretscher. After gaining the approval of Dr White and Dr Bowen in Australia, I accepted. My starting salary was to be £595, which seemed to me a princely sum, though it was appreciably less than the equivalent for a man; it was to take eight years for full equality to be established as a result of an equal pay ruling. But I did not worry about the discrepancy, feeling that I was fortunate to be paid at all to do what I most liked doing.

It was agreed that I should continue at the Cavendish until the manufacture of the magnetic spectrometer was complete; so it was in June 1951 that I joined Harwell.

Chapter 10

Harwell

Shortly before my move to Harwell, Dr Bretscher invited me to visit the Nuclear Physics Division to see what was going on, and to discuss what I should be doing there. The site seemed bigger than ever as I approached it this time. The view from the main gate was pleasing enough. A broad avenue, bordered by young trees and flower beds, led to the main administrative block, an extensive red-brick building with big windows outlined in fresh white paint. But beyond it stretched a bewildering array of heterogeneous buildings dominated by the tall red chimney of the BEPO reactor, and signs of much constructional activity. Facing the road by the gate, hoardings announced the building contractors who were operating on the site, including, in large letters, the name CHIVERS. Some time later, while travelling on a local bus, I overheard the conversation of two ladies sitting behind me. 'Did you notice the "Atomic" as you went by?' asked one of them. 'No,' replied the other. 'I looked out for it, but all I saw was a bloody great jam factory.'

Most of the Nuclear Physics Division was located in one of the four big hangars. Inside, against the south wall, a two-storied string of laboratories and offices had been built, with a railed balcony along the upper level. Dr Bretscher, as I soon discovered, liked to survey his empire from this vantage point and disconcertingly to hail people on the floor below in his penetrating voice. His office was at one end of the balcony; a large room looking across the old RAF runway towards the Downs. He greeted me warmly, seeming much more affable than on the occasion of my oral exam, and conducted me enthusiastically around the hangar, pointing out the salient features.

At one end was the building which housed the Van de Graaff accelerator—a home-made machine based on a design by its American inventor, Professor R J Van de Graaff. The machine was contained within a large cylindrical pressurized steel tank, to prevent electrical breakdowns, and was capable of reaching a much higher voltage than a Cockcroft–Walton accelerator. At the other end of the hangar were a low-energy pile called GLEEP—Britain's first nuclear reactor—and an electron linear accelerator. In the centre stood a small Cockcroft–Walton machine and a block comprising workshops, stores, and labs.

It was agreed that I should join the Van de Graaff Group. Bretscher emphasized that the underlying purpose of the accelerator was to produce fast neutrons (by proton bombardment of suitable targets), to be used for making nuclear reaction measurements relevant to Harwell's reactor design programme. However, he also wished to encourage some basic nuclear physics research, provided it showed promise of original results. He looked at me fiercely as he made this point, giving me the impression that his criteria would be exacting.

Figure 10.1 Sir John Cockcroft (right) with some of his Division Heads (1948). Fifth from left is Dr Egon Bretscher. (Courtesy of Harwell Laboratory)

Bretscher then started to question me about my double-focussing magnetic spectrometer which he knew was now almost complete. 'Who will work on it when you leave?' he asked. I said I didn't know. 'Then why don't you bring it with you here, on loan?' he asked. 'You could carry out the tests on it, and then do some experiments before returning it to the Cavendish.' 'Oh, Mr Shire would certainly not agree to that,' I said emphatically, knowing that Shire had a prejudice against Harwell. But Bretscher wasn't satisfied. 'Frisch is about to-day, somewhere in the Division,' he said. 'He's the Head of Department. Let's ask him.' With a couple of phone calls he located Frisch and asked him to come to his office. Frisch soon appeared and Bretscher put the proposition to him. 'I don't see why not,' said Frisch. 'But Mr Shire would object strongly,' I protested. Frisch just shrugged his shoulders and said that he would speak to Mr Shire about it. 'So that's settled,' said Bretscher gleefully. Frisch went off without further comment, saying that he had to catch a train, and I was left with a sense of foreboding.

Back in the Cavendish I wondered what to do about the situation. I knew that Frisch would take no action, but I couldn't very well let the matter rest for fear of Bretscher's reproof. Confronting Frisch, I asked him if he was going to see Mr Shire. 'Oh, yes,' he said. 'We'd better both go.' Shire looked at us suspiciously as we sat down in his office. Frisch remained silent. I looked at him expectantly. 'You know more about it than I do,' he said to me. 'You'd better explain.' Reluctantly I relayed the conversation which had taken place in Bretscher's office. I expected a strong reaction from Shire, but had not bargained for the intensity of his fury. Characteristically Frisch glanced at his watch and disappeared, leaving me to bear the full brunt of Shire's wrath. He seemed to think that I had planned the whole thing, egged on by Bretscher, whom he obviously disliked intensely. It was unfortunate that I left the Cavendish with an unfavourable feeling towards Mr Shire. He was, I knew, well regarded by other staff members, like Burcham, and the administrative function he loyally fulfilled was obviously a difficult one. Looking back, I realise that it is much to his credit that the Department functioned smoothly during that post-war era.

When I returned to Harwell I told Bretscher what had happened. He nodded. 'That's what I expected,' he said, with a gleam of triumph in his eyes. Evidently he felt that he had paid off an old score as far as Shire was concerned. He's a tough customer, I reflected; I could see that I must learn to stand up to him. Then he proposed that the Division should make its own double-focussing spectrometer. I made it clear that I didn't want to have anything to do with the building of a new one. 'That's all right,' he said. 'The engineers will look after the project if you just hand over your drawings.'

Attached to the Van de Graaff Group was a young man with an engineering background, to whom I offered my drawings of the Cavendish spectrometer. His lips curled at the sight of my amateurish-looking efforts. They had been good enough for the Cambridge Engineering Department, but apparently were not good enough for Harwell. He cast them aside saying that it would be better to start again. Offended, I left him to it. But when, later, he showed me the elaborate professional drawings which the Engineering Services Division had produced, I noticed that the tolerances specified on some of the crucial pole-piece parts were actually overlapping: the pieces, when made, might not fit into each other. He assured me that this had been done deliberately to ensure the greatest precision in their manufacture. I couldn't help feeling some vindictive satisfaction when the parts finally appeared and were found not to fit; they had to be made again. There were further engineering problems before the spectrometer was finally brought into use and the reservations I registered then about the Engineering Services Division persisted for some time.

However, in the Van de Graaff Group I found some able and lively physicists with whom I readily integrated, appreciating their spirit of camaraderie. The three I mostly worked with were from Cambridge. Basil Rose, who had joined Harwell in 1948 from the Canadian Atomic Energy Lab at Chalk River (the successor to Cockcroft's wartime Lab at Montreal), was clearly a man of rare intellectual power; I was not surprised to learn that he had achieved a double first, in physics and mathematics. Roger Hanna and John Newton, both contemporaries of mine at the Cavendish, I already knew as very competent physicists and entertaining companions. I shared an office with John, and found that he had many interests. It was through his being a member of the chess club that I first met a chubby, cheerful-looking young man from the Theoretical Physics Division, who was the club secretary. Occasionally he would knock on our office door and say to me, very politely, 'So sorry to disturb you.' I could not have guessed that this mild-mannered man was destined to create many

Figure 10.2 The Van de Graaff Group 1952, celebrating at a local pub. Front row, first on left: Roger Hanna; first on right: Basil Rose. Second row, centre left: John Newton; centre right: the author. Back row from left, first: Ted Pyrah; second: Doug Allen.

disturbances, as a Director of Harwell, Chairman of the UK Atomic Energy Authority, and then Chairman of the CEGB. His name was Walter (ultimately to become Lord) Marshall.

Our Group Leader, W D (Doug) Allen, an Australian with an Oxford D Phil, who had worked on electromagnetic isotope separators in both America and Harwell, had an inexhaustible supply of energy and humorous anecdotes. He threw himself wholeheartedly into the job of overcoming the teething troubles of the new accelerator, introducing extensive modifications. He was supported by Ted Pyrah, an engineer whose enthusiasm and drive bordered on the fanatic. Doug also carried out some significant measurements on neutron fission. In this work he was joined in 1953 by an energetic and able young Scot, Archie Ferguson, who was to become an important, long-standing member of the Division.

Another asset to the group was Ted Sparrow, a remarkable young man who joined us in 1952. His formal education had come to an untimely end at the age of thirteen when his school was bombed out during the war; but with his innate intelligence and practical aptitude, combined with good nature and flair as an organizer, he became an indispensable member of our group.

Altogether we were an effective group during a period of intensely exciting developments in experimental and theoretical nuclear physics, and I was very happy to be part of it. Although a Research School of Physical Sciences had by then been established under Professor Oliphant in the Australian National University at Canberra, and was giving promise of opportunities for nuclear physics research in my own country, I soon decided that I wanted to stay at Harwell. So my mother, visiting Sydney briefly, resigned from her school and sold the Turramurra house. On her return we bought a house in Abingdon, a pleasant Thames-side town between Harwell and Oxford, and she obtained a post as mathematics mistress at St Helen's School in Abingdon. My transition to Britain now seemed complete. Ironically, in 1970 John Newton became Head of the Nuclear Physics Department in Canberra. When I heard this news I thought of Joe Pawsey's words of encouragement to me when he said that it was in the interests of good Commonwealth relations for English scientists to take posts in Australia while some Australians transferred to England; I felt that John was compensating for the move that I had made.

Most of our Van de Graaff Group experiments were initiated with the objective of obtaining data needed by the reactor designers. However the new techniques we developed frequently suggested further lines of research of interest for fundamental nuclear physics. My most productive work in this period was such a case. I started with Basil Rose to study the inelastic scattering of fast neutrons: the reaction in which a neutron, instead of bouncing off a nucleus like a billiard ball, transferred some of its energy to the nucleus, the latter being left in one of its excited states. It was realised that this process should be investigated to determine its effect on the slowing down of neutrons produced during fission in a reactor core. We observed the reactions by detecting the gamma rays emitted from the nuclear excited states. I soon discovered that the technique not only allowed measurement of the relevant reaction probabilities, but also opened up fresh ways of studying nuclear excited states. When Basil left our Group at the end of 1953, to become Leader of the Cyclotron Group, I carried on with this work with increasing enthusiasm. It happened that two members of the Theoretical Physics Division, Brian Flowers and Phil Elliott, were working on an important model of nuclear structure, called the shell model, from which they were predicting excited states in light nuclei. They urged me to make some measurements on fluorine-19 to compare with their theory. This I was able to do, with some unexpected results: I found more levels than their model predicted; it was fun watching the theorists sort this out. Altogether, inelastic neutron scattering proved to be an effective method of studying nuclear structure.

Like all the Groups in the Research Divisions, the Van de Graaff Group was able to take advantage of the vast range of resources available on site. In a spirit of friendly cooperation we were given innovative equipment, special kinds of targets, radioactive sources, and the benefit of specific skills of other Groups. This cooperative attitude was a particular feature of Harwell, engendered largely, I believe, through Cockcroft's influence. His personality seemed to pervade the whole Establishment. Although the latter was, by 1953, about 3500 strong, with some 750 professionals, Cockcroft appeared to know everyone. He used to come round the labs periodically to talk to us, or, more accurately, invite us individually to talk to him. I remember once giving him a résumé of my current work, pointing to some results pinned up on a board. After talking for about five minutes I stopped. He continued to contemplate my graphs, in complete silence. Feeling that I must do something, I started up again, enlarging on my previous explanations. This time, when I had finished, he nodded, still wearing his benign sphinx-like expression, said 'Very interesting,' and went off. But he gave me the impression of having taken in the essentials, storing them away in his prodigious memory, and I felt encouraged.

Indeed everyone seemed to thrive during that era of Cockcroft's Directorship, and many advances were made: for instance, in 1956 Harwell shared in the triumph of seeing Britain's first nuclear power station opened by the Queen at Calder Hall; great progress was achieved in the fast reactor programme; and ZETA (Zero Energy Thermonuclear Assembly) seemed close to achieving the thermonuclear fusion of deuterium nuclei and opening up bright prospects for the power generators of the future. Harwell's reputation in fundamental and applied research stood high, and I revelled in the general spirit of pride and optimism.

There were some memorable characters among the Harwell staff at that time. Bretscher, with his forthright, at times abrasive, manner and his absent-mindedness, was certainly one of these. His tall, erect figure was a familiar sight as he cycled around the Establishment, though he was sometimes seen returning to his office on foot, because he had forgotten where he had left his bicycle. He was well known for his eccentricities. One of his assistants, Peter Varley, told me of the occasion when Bretscher made some statement to him. 'But a short time ago you told me the exact opposite,' said Varley. 'I did not,' declared Bretscher. 'I have it in writing,' responded Varley. 'Show it to me,' Bretscher demanded. Varley duly produced the relevant paper. Bretscher contemplated it for a moment and then tore the sheet across. 'It is no longer in writing,' he said. Bretscher had little patience with administrators. Once I heard him arguing with one of them over the telephone. 'Remember,' he said peremptorily, 'you are here to serve us.' I wonder how he would have survived under today's conditions.

Another cycling enthusiast who, like Bretscher, lived close to the site, was Dr Eugen Glueckauf, a German refugee and an eccentric but highly regarded chemist. He was a distinctive, not particularly good-looking man, with a shock of curly white hair. His wife once told a friend of mine, with some indignation, how, while riding his bicycle, he had an altercation with a lorry driver. The latter, leaning out of his cab, shouted 'Oi, Einstein! What d'you think you're doing?' 'But you didn't mind his being likened to Einstein, did you?' asked my friend. 'Oh, yes,' said Mrs Glueckauf, 'because Einstein wasn't handsome like my Eugen.' It is interesting to note that the lorry driver had heard of Einstein and apparently had seen a picture of him with his unruly mop of hair. Surprisingly, of all the outstanding scientists over the last century, Einstein, known chiefly for his esoteric mathematical theories of relativity, is the one who has captured the imagination of the general public.

Soon after my arrival at Harwell I met Dr J V Dunworth, my elusive Cavendish Supervisor, who had transferred to Harwell and become Head of the Reactor Division. He was a large, plump man with a round, chubby face and a somewhat indolent manner, by no means devoid of charm. He told me that he had once been mistaken for the flamboyant American pianist and showman, Liberace; stepping from the Continental train at Victoria Station, he was greeted with a red carpet and a bevy of reporters until they discovered their mistake. 'I'll always be able to say that I was once treated as an important person,' he commented. He denied all knowledge of having been appointed my Supervisor at the Cavendish, and I gave him the benefit of the doubt. I enjoyed his quick wit and ready sense of humour in my contacts with him.

There were few women scientists at Harwell. Mrs Katie Glover, who had graduated in physics, but turned to chemistry, was leader of the section in the Chemistry Division which produced standard radioactive sources. She was a well known personality at Harwell, cheerful, helpful, and much in demand for her special skills.

In the Physics Divisions there were three women besides myself. Eleanor Bowey, frustrated in her hope to study physics at a University, was a technical officer in the Linear Accelerator Group. She was said to keep a spare engine under her work bench, in case of sudden need for her little Austin 7. Dr Dorothy Skyrme, a physicist whose husband Tony Skyrme was a member of the Theoretical Physics Division, worked on the cyclotron. She was a pleasant but serious, rather reserved person, so I didn't get to know her well. I was the more intimidated after she and Tony successfully undertook the formidable enterprise of driving, except for two brief sea passages, all the way to and across Australia. They both left Harwell when Tony took a post in Malaya and then a Chair in the University of Birmingham; Dorothy became a part time lecturer in the Physics Department there.

The other woman physicist was Pam Rothwell, a colourful personality who had joined the Cosmic Ray Group while it was headed by Pontecorvo. Small, vivacious, and with an exuberant enthusiasm for life in general and physics in particular, she worked at first on a laboratory experiment; then, having a penchant for outdoor life, she seized a chance to join Terry Price in a cosmic ray experiment to be carried out in the French Alps. There are many stories of Pam's escapades, for example the occasion on which, during a violent argument with Terry over the right to a sheet of data, she stuffed the paper inside her dress. He then blocked her exit from the room; so she climbed out of the window, which was one floor above ground, and, using successive window ledges as handholds, made a traverse across the building, to the astonishment of the Group Leader next door, until she reached the window of her own room.

In spite of madcap tendencies, Pam had the qualities which I recognize as important for a successful woman physicist: a deep interest in her subject, an independent spirit, and a determination to succeed. She took in her stride marriage and the bringing up of a daughter, and ultimately became a

Senior Lecturer at the University of Southampton. There she organized and led an upper atmosphere research group, satisfying her love of travel with observational work in various parts of the world, including a remote base in Northern Scandinavia. She took a great interest in the succession of PhD students who passed through her hands, relishing the satisfaction of seeing them successfully launched into good physics careers.

After Pam and Dorothy Skyrme left, I was for most of the time the only woman physicist. I rather enjoyed my unique status, and never felt that my sex was an obstacle to my professional progress. Once, while still quite a junior staff member, I found an opportunity of turning my sex to advantage. A scientific manufacturing firm had failed to deliver on time a piece of apparatus that I had ordered. There were technical difficulties, I was told when I enquired by telephone. Somewhat suspicious, I asked to speak to the person concerned with making the device. 'This is Dr Freeman's office at Harwell,' I said, when he came on the line; I pressed him to explain the technical aspects of the problem. He was obviously taken aback by the extent of my scientific knowledge and jumped to the conclusion that if Dr Freeman's secretary was as competent as this he must be a very important person. I received my order within a few days.

I did, however, find one small avenue effectively out of bounds to me. I had a much treasured Morris Minor, which I used when I knew I was going to be working late. But one day, having come to work on a Harwell bus, I unexpectedly found myself running on the Van de Graaff in the evening; I decided to go home on one of the ten o'clock buses which were laid on for the regular shift workers. There were about half-a-dozen buses at the assembly point, and I asked the Supervisor which one went to Abingdon. He looked at me very suspiciously. 'Are you a genuine worker?' he asked. 'Only genuine workers are allowed on these buses.' I assured him that I was bona fide, so he grudgingly let me on the bus, which was practically empty to start with; I tucked myself into a seat near the back. Soon a large group of men got in, chattering and laughing noisily amongst themselves. Then one of them suddenly spotted me and exclaimed loudly 'Cripes, there's a *lidy* on the bus.' A deathly hush ensued. I was evidently an unwelcome and inhibiting intruder upon their conversational freedom. I never tried that mode of transport again.

Always a particular pleasure for me has been that of attending nuclear physics conferences. My first was a memorable five-day international meeting in 1953 at the University of Birmingham. It was a great thrill to meet leading nuclear physicists from other countries and to hear at first hand of dramatic new experimental and theoretical developments. Over the years I have been privileged to attend many such conferences in various parts of the world. I have always revelled in the stimulus of scientific exchanges, the morale-boosting effect of presenting my own work, and the feeling of being part of this special international community, with common interests transcending social and political boundaries. Being a woman was, I think, an advantage to me on these occasions. Since there were always very few women present, I soon became better known than would have been the case had I been just another man; I was never short of interesting company during such meetings.

In 1956 a small conference was convened at Harwell by Dr Bretscher for the exchange, with groups in the United States and Canada, of data on neutron interactions in reactors. At my instigation a session on the new, as yet not fully recognized, subject of inelastic neutron scattering was added to the topics for discussion. I had the great satisfaction of opening this session with a review paper in which I was able to draw together into a logical whole the separate known facts on the subject. The Americans in particular liked my talk, and the result was that, in the following year, they invited me to present a further review paper at an International Conference to be held at Columbia University, New York. This adventure will be described in the next chapter.

In the second half of the decade, the Van de Graaff Group acquired three valuable new members. First, in 1955, came a capable, cheerful, outgoing Canadian called John Montague, always going out of his way to be helpful. In the following year we gained Arnold Jones, a talented, down-to-earth ex-Cavendish Welshman, who kept us entertained with his characteristic, somewhat cynical brand of humour. At the same time Eric Paul, whom I remembered as John Jelley's Cavendish colleague, joined us from the Chalk River Laboratory, where he had been working on their Van de Graaff accelerator.

Meantime a new development, which was to be particularly important to me, was being considered for our Group. News had arrived from America of an innovation in Van de Graaff accelerator design: the tandem Van de Graaff, which was expected to double the attainable energy of a proton beam. It had two accelerator tubes, with the positive high-voltage terminal in the middle. Negative ions (protons with two electrons attached), injected into the first tube, were accelerated to the terminal, where the electrons were stripped off, and the protons were then accelerated further in the second tube. The first machine of this type was to be built by the Massachusetts firm, High Voltage Engineering Corporation (HVEC), for installation at the Chalk River Laboratory. Doug Allen, having gained much experience with our Van de Graaff, envisaged the design and building, with the aid of a British firm, of a tandem accelerator for Harwell. His enthusiasm for the venture was reinforced by Eric Paul when the latter arrived with the latest news of the Chalk River plans.

It was a bold engineering enterprise to contemplate, and the initial stages were fraught with complications because of the plan for a tandem simultaneously conceived at the Atomic Weapons Research Establishment at Aldermaston. This, though independent of Harwell, was part of the same recently formed United Kingdom Atomic Energy Authority (UKAEA). The Aldermaston project was master-minded by K W Allen, who had been at the Cavendish when I arrived there; he was ambitious to build up a nuclear physics research group similar to ours at Harwell. He was a formidable opponent when it came to competing with W D Allen for UKAEA approval of the costly tandem project. In the end agreement was obtained to have two accelerators, one in each Establishment, the main justification in each case resting on the proclaimed necessity of higher energies to produce neutrons for measurements relevant respectively to the atomic energy and atomic weapons programmes. In a somewhat uneasy alliance, responsibility for the development of the various components of the machines was shared between the two Establishments.

Doug left our Group, with Ted Pyrah, transferring to the Accelerator Division so that he could devote himself full time to the tandem enterprise, and Eric Paul took over as Group Leader. Maintenance and development of the Van de Graaff became the responsibility of Frank Pilling and Sid Waring, both quiet, eminently capable individualists, supported by a technical designer, Ken Knox, a bouncy Liverpudlian. With the ever-ready help of Ted Sparrow, they formed a remarkably successful team. Eric applied himself with the same intensity that I had seen in him at the Cavendish, making original contributions to the research, and laying ambitious plans for the future organization and development of the Group. As well as the tandem, he envisaged having a small pulsed Van de Graaff accelerator from HVEC, dedicated to neutron work. This came into being in 1960 and the Group, by then with three accelerators, was renamed the High Voltage Laboratories.

By now the reader may well be wondering what of John Jelley. Not long after my arrival in Harwell his father died and he had considerable problems with his demanding mother. However, we enjoyed each other's company when opportunities arose, and I followed with interest his research activities in the Cosmic Ray Group. He was able to do some very original work while satisfying his zest for the great outdoors. He first discovered that cosmic ray particles—the

Figure 10.3 Some of the staff of the High Voltage Laboratories, 1960. Front row from the left: John Montague, Dr Bretscher, Eric Paul, Sid Waring, Archie Ferguson, John Fruin, Frank Pilling. Second row, second from the left: Brian Hooton; third from left: Ken Knox. Back row, first from right, Ted Sparrow; third from right, Arnold Jones. (Courtesy of Harwell Laboratory)

high-energy radiation continuously showering the earth—
produced pulses of light when they passed through a column
of water. This light he identified with a type of radiation
named after the Russian scientist Cherenkov. He then assembled a simple detector consisting of a mirror and a
photomultiplier mounted inside a dustbin, and set it up
outside a small hut near the boundary of the Harwell site, to
see if he could on dark nights detect Cherenkov light flashes
resulting from cosmic ray particles passing through the atmosphere. The experiment was a great success and led to a series
of definitive studies which he carried out with a colleague, W
Galbraith, at Harwell and at the Pic du Midi Observatory,
high up in the Pyrenees. The now flourishing field of gamma
ray astronomy developed directly from this work.

John revelled in his night work and the remoteness of his
Harwell hut, from which he could observe the local wildlife,
including a family of little owls who nested underneath the
building. He seemed quite impervious to the cold, and even in
midwinter, when my instinct was to curl up and hibernate, he
went about without an overcoat. The man behind the counter
in the little post-office adjoining the Harwell site once remarked to a friend of ours that some Harwell employees must
be very poorly paid because he had noticed that one chap, who
came in regularly, couldn't afford an overcoat. Our friend
immediately knew whom he meant.

Later, John agreed to undertake a project to collect a series
of samples of the gas at the top of the earth's atmosphere,
some 100,000 feet up, for radioactivity analysis. This was
achieved by sending up huge balloons, with the cooperation of
the Bristol University Cosmic Ray Group; each balloon carried
apparatus in which, by remote control, a valve could be
opened to allow a gas sample at the ceiling height to be frozen
onto a liquid-nitrogen cooled collector; the valve was closed
again after a few hours, and then the apparatus was released
from the balloon so that it parachuted back to earth. John and
his team had to rush off in a van to pick up the equipment
wherever it landed. Meteorological forecasts of wind strength
and direction were not as reliable in those pre-satellite days
as they are now, so that chasing after the quarry led to some
quite wild adventures. The help of local farmers had often to
be invoked, and their reactions to the arrival of this myster-
ious object apparently from outer space were often hilarious.
Once, the balloon had to be shot down because it was
threatening to enter an airline corridor; on another occasion

Figure 10.4 John Jelley and the author on their wedding day, February 1958.

the apparatus tantalizingly dropped into the sea just off the Welsh coast; and once it landed on the greenhouse of a stately home and finished up among the tomato plants. The lady of the house was none too pleased, and at first refused to allow this curious looking group of young men to retrieve their gear. Finally she relented, with the admonishment 'Don't let it happen again.' John enjoyed practical challenges of this kind, though he was somewhat frustrated because the project was not concerned with fundamental research. He was glad to get back to this when he began to pioneer the detection of cosmic ray showers by radiowave observations.

Meantime, early in 1958, John and I finally decided to get married and to have what proved to be like an extended honeymoon: a sabbatical year in the United States. Exchanges of this type with American and European research centres were encouraged by Harwell in those halcyon days.

Chapter 11

American Interludes

My first trip to the USA was in 1957, to the Neutron Interactions Conference at Columbia University. The transatlantic flight, in that pre-jet era, was a noisy and bumpy ride taking some thirteen hours, but the four of us from Harwell were buoyed up by the excitement of this new experience. New York gave me an unforgettable first impression, with its unique atmosphere, its skyscrapers—even taller than I had imagined—and its geometrical layout of streets and avenues tapering like canyons to distant skylines.

Accommodation had been reserved for us at the New Yorker Hotel. It rose skywards for some 43 stories, and its busy, lavishly appointed lobby was vast. 'I've come for the Conference,' I volunteered to the man at the front desk who was checking me in. He shot a momentary, uninterested glance towards a large noticeboard which announced the conferences for the day and the particular hotel suites to which they had been allocated. The first of several was a Dog Food Convention. Needless to say, there was no mention of our meeting, since it was not being held in the hotel. Blushing at my own naivety, I followed a porter up to my room on the nineteenth floor. Here was unaccustomed luxury, including a private bathroom and plumbing not only for hot and cold, but also for iced water. The latter was a welcome novelty, as the weather was decidedly hot, though I soon discovered that Americans had an obsession for icing their cold drinks to such an extent that there was no taste left in them.

I rejoined my Harwell companions, and we set out with maps and a subway guide to find our way to Columbia University. I had been led to believe that New Yorkers were unfriendly, but we soon encountered a man with a strong New

York accent so anxious to help some obvious strangers that we found him hard to get rid of. Later, when I had occasion to take a taxi, the driver, learning that I had come from England, assumed that I must be poor and gave me back the money that I had offered him as a tip.

We reached the conference venue in time for the initial reception, and I soon found myself being greeted warmly by people that I had met at previous meetings, and, through them, making new contacts. In the relaxed and friendly atmosphere, my spirits rose and my self-confidence expanded. Altogether the conference was a great success for me, and my talk invoked satisfying comments.

Before returning home, I visited some Labs including the Oak Ridge National Laboratory in Tennessee: a large establishment somewhat similar to Harwell. I flew with a couple of my Harwell colleagues to the nearest airport, and we then took the airport limousine to Oak Ridge. Sharing the transport with us was a very pleasant man who chatted freely, and had soon learned where we were from and what we were doing. 'Do you work at the Oak Ridge Lab?' I asked. He said he did. 'Are you also a research physicist?' I then enquired. 'Well, no,' he said, in an apologetic tone of voice, 'I'm an administrator, in fact.' It was not until we had reached our destination and he had gone off, wishing us a pleasant and profitable stay at Oak Ridge, that I learned that he was the Director of the Laboratory. Though feeling slightly embarrassed as I realised that I had probably given the impression of regarding administrators as inferior to research scientists, I appreciated the easy-going informality and lack of class-consciousness in this important man; I was beginning to recognize these characteristics as typically American.

In the following year, having decided to spend a sabbatical in the United States, John and I had the problem of finding suitable places for both of us in the same locality. In the end we had a choice between two possibilities. One was to go to the California Institute of Technology in Pasadena. I was offered a nuclear physics Research Fellowship there by Professor Willie Fowler, a brilliant physicist and a delightful person whom we had got to know when he spent a year in Cambridge. Willie proposed that John should join the Radio-astronomy Group. This appealed to John, the only snag being that the radio telescope was a few hundred miles from Pasadena.

The alternative was Cambridge, Massachusetts. I had the

offer of a Research Fellowship at MIT, the Massachusetts Institute of Technology, in Professor W W (Bill) Buechner's High Voltage Laboratory. I had not met Professor Buechner but knew of his high reputation in the nuclear physics community. His Lab, the academic home of Professor R J Van de Graaff, contained the biggest Van de Graaff accelerator in the world, together with a unique magnetic spectrograph which he had devised. Significant new data on nuclear excited states were flowing steadily from his group. At the same time John was offered a research post at the Harvard College Observatory, which was only two or three miles from MIT. The proposal was made by Tom Gold, a Harvard Professor whom John had first met in Cambridge. Tom, an Austrian refugee who boasted that in England, early in the war, he had been transferred straight from internment to a top-secret job, was an extrovert of outstanding intellectual ability with unbounded enthusiasm for physics, and a prolific producer of original ideas. He suggested that John develop an ultra-sensitive detector for radioastronomy, using a solid-state maser (forerunner of the laser) recently invented at Harvard by Professor Bloembergen.

Mainly on the grounds of geographical convenience, we chose the MIT/Harvard option and, crossing the Atlantic luxuriously by boat, we arrived in Cambridge early in October 1958. We stayed first at the Brattle Inn near Harvard Square. One of the attractive colonial-style wooden houses, of which there were many in the area, it looked particularly ancient. Inside, the floor of our room was on such a slope that the head of the bed had to be propped up on six-inch thick wooden blocks. Altogether, the centre of Cambridge gave a surprising impression of antiquity, quite apart from the handsome buildings of the Harvard College campus which adjoined the Square. About the local people too, there was an air of old-fashioned courtesy and adherence to traditional social customs, which we later found to be characteristic of the whole New England area.

A short ride in a trolley-bus from Harvard Square brought me to the arresting sight of MIT's main entrance with its broad steps and row of tall, slender columns topped by a big white dome. The clean and simple principles of the design were repeated in the main buildings set around spacious quadrangles, to give an impression of classic elegance and quiet dignity. I felt that the architecture somehow reflected the purposes of the Institute: to promote education and

Figure 11.1 The Maclaurin Building, Massachusetts Institute of Technology. (Courtesy of MIT Museum)

research at the highest level in science and engineering, while recognizing the value of, and imposing on the undergraduate curriculum, some studies and appreciation of history and the arts.

Inside, there was a busy, inviting atmosphere. But first I had to go through the admission procedures. There are many things I admire about American life, but ponderous and sometimes illogical bureaucracy, when it rears its head, is not one of them. Though I was only taking a temporary job, I was given a great sheaf of forms to fill in, demanding endless personal details; many of them were repetitive, and one, a wartime relic I imagined, stated at the bottom of it that if I could not write I must put a cross where the signature was required, in the presence of a witness. Then I had to go through a thorough medical examination; the report made no adverse comment apart from recording that I had flat feet.

At last I was ready to set out for the High Voltage Laboratory which was in MIT's unattractive back-yard of more recent developments; it looked a bit like part of a factory. Inside, the huge Van de Graaff accelerator, with its associated equipment and target room, occupied most of the building. The labs and office space were cramped, but there was a cosy, welcoming, atmosphere, and I was touched to note a desk on which had been set a nicely finished wooden block bearing a name plate which read: DR JOAN FREEMAN. Professor Bill Buechner greeted me warmly. He was a man of

Figure 11.2 Professor William Buechner (c 1964). (Courtesy of MIT Museum)

few words, but with an open, cheerful disposition. He obviously ran the Lab with a firm hand, but I could see that he was held in affection as well as respect by everyone there. His sociable and helpful secretary, Mrs Mary White, who was as talkative as he was uncommunicative, was located a short walk away in another building, the old High Voltage Laboratory, where she had worked for Professor Van de Graaff before he was forced to retire through ill health. Van de Graaff came in occasionally, and I had the opportunity of meeting him. I was captivated by his old-world charm and gentleness.

The research group in Buechner's lab comprised one other temporary Fellow, a quiet, hard-working New Zealander called Bob White (who appears again later in my story), some MIT graduate students, several visitors, and two resident staff members: Associate Professor Harald Enge, a handsome, able, and likeable Norwegian, who was to make a name for himself in original magnetic spectrometer design, and Anthony Sperduto (known always as Spud), of Italian parentage and temperament. He had joined the Lab after getting a bachelor's degree, and Buechner, recognizing Spud's natural talents, had

Figure 11.3 Anthony Sperduto at a lab party. (Courtesy of Mrs Hester Sperduto)

encouraged him not only to look after the accelerator, which he did with skill and resourcefulness, but also to participate in the experimental work; Spud's name appeared on many of the publications. He was warm-hearted, unstintingly helpful to everyone, particularly the students, and had an endearing and entertaining personality. Although meticulous when it came to the accelerator, he was haphazard about the care of his own possessions: at one stage, when the accelerator pedal of his rickety old car had broken, he was operating the throttle with a piece of string looped over one finger. He seemed frequently to be involved in extraneous activities of one sort or another, and would sometimes, if his presence wasn't vital, creep furtively into the lab half-way through the morning, muttering something about a job "for the Chinese Government". I became very fond of Spud, and grateful for the friendship and support he gave me.

The project Buechner proposed for me was that of making a gas target with which I could measure nuclear reactions leading to the excited states of neon isotopes. He pointed out that the Lab had collected data on many nuclei which could be studied with solid targets, but there was a significant gap at neon because this was a noble gas, existing only in gaseous form. I could see that there would be problems in making adequately thin windows for the gas cell, and in handling the separated isotopes of neon, which would be available only in very small quantities. But it seemed an interesting and worthwhile project which, hopefully, I could complete in a year, so I readily agreed.

Exchanging notes later with John, I learned that his project seemed considerably more formidable than mine, but he was excited about the opportunity of exploiting the new maser phenomenon. 'Anyway, I won't be on my own in this,' he said. 'Tom Gold has hired another man to work with me. In fact he's an Australian on leave from the Radiophysics Lab. You may have heard of him; his name is Brian Cooper.' 'Brian Cooper!' I exclaimed, astonished at this incredible coincidence. 'My old friend and contemporary at Sydney University and RP!' In great excitement I made contact as soon as possible. There was Brian, as large as life, giving his characteristic slight jerk of the head, but otherwise not displaying his surprise. Yes, Pat was with him, plus their young family. It was a thrill to see them again after twelve years, and to extract from Brian some of the recent RP gossip, while Pat stirred in me a sense of nostalgia as she talked about Sydney. I much enjoyed their company through the year, and John and Brian worked well together, their effective partnership contributing largely to their ultimate success.

After some searching, John and I found an ideal apartment, within walking distance of Harvard Square. It looked out onto huge trees frequented by squirrels and brightly-coloured blue jays with their raucous calls. Inside, the apartment was comfortably and attractively furnished and had the spaciousness typical of American homes. 'There are four closets,' announced the owner when he was showing us round. Thinking of water-closets, I expressed surprise. He opened a door along the passage and displayed a linen cupboard. This was one of many examples of the confusion and amusement we found in unfamiliar American word usage. It took me weeks to discover that cornflour was called cornstarch, and when I asked for cotton wool at a drugstore, the assistant said 'Make up your mind. Do you want cotton or do you want wool?' One day as I arrived at the MIT Lab, Spud said to me 'While you were out, someone called and asked for you.' 'What did they look like?' I asked. Laughing uproariously he said that they didn't have video telephones yet. As for the American equivalents of the English word "tap", we both got into difficulties over the alternatives of "cock", "valve" and "faucet".

There were many pleasurable aspects of life in Cambridge. I recall particularly the warmth of American-style hospitality, and the opportunities we seized to explore the New England area, with its distinctive architecture, history, and scenery. The varied wildlife we found of absorbing interest: I was quite carried away by the charm of chipmunks and of birds like the dapper little chickadees. Once, seeking a place to stay in a delectable part of Cape Cod, I knocked on the door of a house which offered holiday cabins. Just as a man opened the door, an unfamiliar bird alighted on the grass quite close to me. I was so fascinated by it that, instead of saying that I wanted accommodation, I just pointed at the bird and asked the man what it was. He must have thought I was a loony because he promptly slammed the door and I beat a hasty retreat, fearing that he might set the police on me.

At MIT I found much to interest me in addition to the intimate little world of the High Voltage Lab. In the main physics building there were stimulating colloquia to attend, and a number of lively research groups to visit. I got to know the Cosmic Ray Group in particular, since this was also of interest to John, and was impressed with how hard they worked. They were on the sixth floor, and occasionally, around lunch-time I would meet one of them in the elevator. He would go down to the basement to "grab a sandwich" from the machine there, and by the time he was back at his own level he had finished his lunch. I wondered, though, if the research students were not so hard pressed with their course work that they had no time to stop and think. Chatting to one of them I happened to mention that my married name was Jelley. 'Are you related to the Jelley who wrote the well-known book on Cherenkov radiation?' he asked with interest. 'He's my husband,' I replied. 'You mean he's still alive?' was the incredulous response. Evidently it hadn't occurred to him that the authors of his text-books might not all be dead. I gained the general impression that the courses at both undergraduate and graduate level tended to be too intensive. I once asked one of the High Voltage Lab research students a nuclear physics question. 'That's on page 874 of Evans (the standard nuclear physics textbook),' he said without hesitation. Then I asked him another question. He looked blank and uninterested. 'That's not in Evans,' he replied, dismissively.

A highlight of my period at MIT was the presence of Professor Victor Weisskopf, a brilliant exponent of nuclear theory and creator of original ideas, who made a deep impression on me. It was lucky for me that he was there in 1959 as he had been away on sabbatical during the previous year, and was to go in 1960 to the European Centre for Nuclear Research (CERN), in Geneva, to become Director-General for four years. He was an Austrian by birth, one of the many outstanding scientists whom the USA acquired as a result of Nazi anti-semitism; a big man, of distinctive appearance, with dark thinning hair, bushy black eyebrows, and a sensitive face. His strong, outgoing, warmly human personality seemed to embrace the whole Physics Department, in which he took a lively interest. Even with me, just a temporary visitor, he interacted in a kindly and stimulating way, enquiring about my activities and taking note of my impressions of the American scene, as well as imparting his own views.

Weisskopf was an excellent lecturer, and the informal talks he liked to give, especially for the benefit of students, were a particular delight to me. Once, while he was describing some recent discovery of great significance for astrophysics, I remember his breaking off spontaneously to remark on the sense of pleasure and awe he derived from gaining a new insight like this into what he called one of the miracles of the natural world. These miracles, he said, were for him on the same level as those he found in artistic creations such as

Figure 11.4 Professor Victor Weisskopf. (Courtesy of MIT Museum)

music, painting and sculpture. More than any other scientist I have met, Weisskopf gave tangible expression to my intrinsic feeling for the exhilaration and satisfaction to be gained from having some understanding of physics.

Once, after hearing him give a superb public lecture, I asked him whether, after so many successful talks, he still felt nervous when he first got up to speak. 'Oh, yes,' he replied, 'I don't believe that anyone can give a good talk unless the adrenalin flows a little to begin with.' That may be a necessary condition, I thought to myself, but it's certainly not sufficient for reaching Weisskopf's standard.

Weisskopf had a great sense of humour, too, and usually included some anecdotes in his lectures. I recall one story he told of Pauli, a famous theorist under whom he had worked at Zurich. Pauli was giving a lecture with a mathematical argument which he set out on the black-board. At one stage he said 'It is obvious that . . .' and wrote up another equation. Standing back from the board, he contemplated it in silence; then he walked out of the room. Five minutes later he returned. 'Yes, it is obvious,' he said.

A few years later I saw an example of Weisskopf's sense of fun in a letter which appeared in the American Physical Society's Journal, *Physics Today*. It was about a wager made between himself and Professor Richard Feynman, a famous theoretical physicist equally well known as a joker. Feynman was to lose the wager if at any time in the next ten years he held a "responsible position". This was defined as a position which "compels the holder to issue instructions to other persons to carry out certain acts, notwithstanding the fact that the holder has no understanding whatsoever of that which he is instructing the aforesaid persons to accomplish". Ten years later Weisskopf reported in the same medium that he had lost the wager.

Soon after my arrival at MIT I received an exciting letter from Dr K K Darrow, Secretary of the American Physical Society: he had heard and enjoyed my talk at the 1957 Columbia Conference, and would I give a 30 minute invited paper at the annual APS meeting to be held in New York in January? This, I knew, was the biggest national physics conference of the year. There would be hundreds of delegates and hundreds of contributed 10 minute papers, with just a handful of invited papers (there were in fact to be just six on nuclear physics). It was an honour to be invited, and by

Darrow himself, whose work at the Bell Telephone Labs on electricity in gases I had known since my Sydney University days. I accepted with a glow of satisfaction.

The meeting was held in the New Yorker Hotel, using all their conference halls, and, this time, featuring prominently on their notice board. The session which included my talk took place in the Grand Ballroom. Standing up to speak, I found myself confronting a great sea of faces extending into the depths of the long hall. This was far bigger than any audience I had addressed before. Weak at the knees, I plunged into my story of inelastic neutron scattering. When the lights were dimmed, so that I could show my slides, it felt more comfortable, and evidently I made a good job of the talk since, after it, Darrow sprang up and praised not only the subject matter, but also my lecturing technique, of which, so he said, many speakers would do well to take note. I felt embarrassed, but found that my American friends took it in good part, saying that lecturing style was one of Darrow's hobby horses.

There were very few women at the APS meeting. In fact the proportion of women physicists in America seemed to be just as small as in England or Australia. At MIT, which was oriented towards science and engineering, only 2 per cent of all students were female. This apparently caused social difficulties since I saw in a booklet "Guide to Graduate Life at MIT" a section entitled "how to meet women". But I found that the handful of women in nuclear physics whom I managed to meet in America were outstandingly able, and highly respected by their male colleagues.

One of these I got to know well. This was Janet Guernsey, whom I first met at the 1957 Columbia Conference. She was a resolute, very likeable person, with an easy, unassuming manner. Of her practical capabilities I felt convinced when I saw her park, in one quick sweep, in a space barely a foot longer than her car. At MIT I learned the impressive story of her career. Having a bent towards the physical sciences from an early age, she majored in physics in 1935 at Wellesley, the highly regarded women's College near Cambridge, Mass. In the following year she married William Guernsey, a historian, and, over a period of ten years, produced five children. In 1943 she seized the opportunity of replacing a Wellesley faculty member who had broken her leg, and then, sixteen years after taking her first degree, she enrolled as a graduate student in experimental nuclear physics at MIT, gaining her PhD in 1955. She worked on inelastic neutron scattering, publishing

Figure 11.5 Professor Janet Guernsey with two of her
students. (Courtesy of Wellesley College)

several papers; it was through this work that I first got to
know her, and I marvelled at her success after such a long
interval away from a laboratory.

In Janet's subsequent career at Wellesley she rose to become Professor and Head of the Physics Department, as well
as taking an active part in furthering the objectives of the
American Association of Physics Teachers of which she was
elected President in 1975. In Janet I saw another example of
determination, an independent spirit, and a real enthusiasm
for her subject—the important characteristics of a successful
woman physicist.

In due course my work in the High Voltage Lab reached the
stage where I was ready to run my first experiment on the
accelerator. I had managed to make a gas cell which survived
all its tests in the target chamber, and I had acquired from
Oak Ridge, at frightening expense, two little glass tubes
containing separated samples of the two main neon isotopes.
With Spud's help everything was set up for the machine run

and one of the neon samples was attached to my vacuum system. At the appropriate moment, sweating with apprehension, I broke the internal seal, allowing the gas to expand into the target cell. If this leaked during the beam bombardment, I knew that the precious gas would be irretrievably lost. The suspense was agonizing because the nuclear reaction products passing through the magnetic spectrometer were recorded on a long photographic plate: this could not be examined until the run was completed and the plate processed. Then at last I could look for particle tracks, under a microscope. With sinking heart I saw nothing. But Spud, knowing better what to expect, soon started to find evidence for particle groups. Before long it was clear that the experiment had worked, and I could look forward to analysing the results after the plate scanning team had carried out their tedious quantitative track counting. Bill Buechner came and peered through the microscope at the photographic plate. After a while he looked up at me, smiling, and, with his usual economy of words, simply said, 'You're a success.' I reflected once again that there was nothing like experimental physics for creating moments of sheer delight.

In the fall, the High Voltage Lab acquired a new recruit. This was Audrey Blin-Stoyle, a very pretty and vivacious young Englishwoman who took a temporary job as a computing assistant to Buechner while her husband Roger was spending a sabbatical year in the Theoretical Physics Department. Her arrival was the first link in a chain of events which was to lead to the most successful research project of my career, as will be recounted in the next chapter.

Audrey was a delightfully warm and friendly person. One day she invited John and me to dinner at their rented house and introduced us to Roger. I was immediately captivated by this handsome young man with his cultured English voice and gentle manner. I discovered that he was a Lecturer in the Theoretical Physics Department at Oxford. I felt surprised that I had not heard of him before, but supposed it was because of his quiet, modest disposition. He told me that he was working on some fundamental aspects of the theory of nuclear beta decay and nuclear forces. It all sounded somewhat beyond my powers to grasp. Little did I imagine that before long I would be struggling hard to understand it.

Towards the end of our period in Cambridge, Mass, which had been extended by three months, I saw Eric Paul at a New York meeting. He brought me up to date with progress at

Harwell, telling me, with satisfaction, that the home-made tandem accelerator had produced its first beam only three months after the HVEC machine at Chalk River reached this stage. The small pulsed Van de Graaff, which he had named IBIS, would soon be operating. Then Eric gave me the news that Dr Bretscher had asked him to take on the function of Deputy Division Head. He would, he said, be maintaining overall administrative responsibility for the High Voltage Labs, but it was necessary for Group Leaders to be appointed to look after the three accelerators, their users, and the support staff. The proposal was that Arnold Jones should run the five million volt Van de Graaff; Archie Ferguson, the IBIS machine; and I should take on the tandem. I was flabbergasted. To be given responsibility for Harwell's newest, very expensive acquisition, and for all the staff involved, was certainly an unexpected turn of events.

Chapter 12

From B to Z

The tandem building was a conspicuous Harwell landmark. Approaching it for the first time, early in January 1960, I asked myself could I really be in charge of this? The 90 foot high, partly concrete, partly aluminium-clad tower, shining in the thin winter sunshine, looked impressively tall—the more so as the rest of the building was single-storied. The laboratories and offices were strung out in a long line with a central entrance facing north towards the Nuclear Physics Division hangar across a broad stretch of grass and newly planted trees. On the south side, the building was in the shape of a large semicircle. This was the target area; it contained the experimental rigs into which the accelerated beam could be directed from a rotatable analysing magnet located beneath the massive steel tank which housed the tandem machine.

The hub of the lab's activities was the control room, a pleasing and spacious area, well lit by daylight from large skylights. Here I found Doug Allen and Ted Pyrah dashing about in their characteristic fashion. Doug took me up in the lift to the top of the accelerator, where the ion source was located in a circular bay with windows giving a fine view of the site. We descended by a spiral staircase from which I could see some large ports in the side of the tank, opened to reveal parts of the accelerator inside: a formidable-looking monster. Doug told me with pride of the good progress that they had made with the machine, though there were still some faults to be ironed out. I had a great deal to learn, I realised.

I talked also to Ralph Dawton who was responsible for the design and building of the all-important ion source. He was a delightful character, tall, thin, and wiry, with a dedicated

Figure 12.1 The Tandem Accelerator Building. The cylindrical section of the tower houses the accelerator; the rectangular section behind contains the lift-well. (Courtesy of Harwell Laboratory)

enthusiasm for his job and a great sense of humour. He was another fascinating example of innate ability for physics triumphing over inadequate early training. He told me once that at his London County Council school he was taught mainly about the British Empire. Later, when he became an apprentice mechanic at the Royal Institution, he decided that he should try to gain matriculation by going to night school, but he had great difficulty in learning the required foreign language, since the French textbook used words like "adjective", "noun" and "verb" which meant nothing to him. Eventually, having been promoted to the position of lab boy, he got his B Sc. When World War II broke out he was sent to Birmingham University to work on microwave radar under Professor Oliphant. There he gained his Ph D. He followed Oliphant to the USA, worked on electromagnetic separators, and then joined Harwell in 1946. As with other successful scientists I have met who had to struggle against unfavourable odds in their early life, he was a particularly tolerant, helpful, and likeable person. He was also very gifted. The skill and originality he brought to bear not only on ion source design but also on beam optics and many mechanical and electrical aspects of the accelerator were to prove invaluable to us.

I soon discovered that all was not running smoothly as far as relations between the hangar and the tandem building were concerned. To begin with, Doug Allen and Eric Paul had fallen out to such an extent that they were communicating with each other by correspondence only. From my point of view, though, the most serious difficulty concerned the tandem technical staff. This problem was to persist long after Doug had formally handed over the tandem to the Nuclear Physics Division and had departed for his next job—to build an accelerator for Denys Wilkinson at Oxford University. It arose because Eric had chosen to draw on the pool of staff available from the Engineering Services Division to provide the operation and maintenance team for the tandem. I had been feeling apprehensive at the thought of being in charge of this unknown and, in my imagination, intimidating group of a dozen or so men, wondering what they would make of having a small, mild-mannered woman as their boss. But in the event my problem was quite different from this.

Pilling, Waring, and the Van de Graaff operating team, responsible also for the new IBIS machine, resented the fact that a different group, belonging to another Division, and with no previous experience of electrostatic accelerators, had been brought in to run the tandem; moreover, the tandem operators were to be paid more than theirs. As a result of this hostile reaction the new tandem staff were very much on the defensive. Peter Humphries, the engineer in charge, was a pleasant, youthful-looking man, small in stature, and cautious by nature, but quietly determined. I found him easy to get on with, and likewise most of his team. But I had to walk a tight-rope in giving them necessary encouragement and, at the same time, maintaining friendly relations with Pilling and Co. Thankfully, I gained support from the scientific staff: John Montague in particular, with his cheerful bonhomie and his considerable practical capabilities, was a tower of strength.

But there were occasional ructions, as when Dr Bretscher suddenly decided to intervene personally, coming over to the tandem building and declaring that Pilling should be put in charge of the accelerator. I wrote him a fierce note asking him not to interfere as this caused serious disturbance to the technical staff without doing any good. He was very annoyed with me about this—not so much because I had ticked him off, he told me, but because I had marked the envelope "Confidential" rather than "Personal". This had meant that his secretary opened the letter and he did not like her discovering that he was being reprimanded by a member of his staff. But he soon got over his sulks, and subsequently treated me with increased respect. I often wondered what his personal feelings were about women in the lab, particularly as he came from Switzerland where women didn't even have the vote, but I never detected a sign of sex discrimination from him. He admired intellectual ability and was very proud of his wife, who had been a mathematician at the ETH in Zurich, where they met. He liked to relate how, after they married and she left the Department, his boss asked him where Hanni was. 'She's looking after the home now,' he replied. 'But she is a better mathematician than you are,' his boss said. 'She should be here and you should be at home.'

The initial disturbances gradually subsided and with relief I saw the accelerator settle down to reliable operation, allowing a variety of interesting experiments to be carried out. The time was shared between High Voltage Lab members, other Harwell groups, and visitors from many Universities. In our hey-day, we had as many as eighty people per year running experiments on the tandem.

Early on I initiated an experimental programme of my own which was to evolve in a remarkable way. While wondering at first what to do, I talked to David West, a quiet, reserved, very able member of the by then disbanded Cosmic Ray Group. He had built a unique detector—a very large ionization chamber filled with helium-3 and krypton gases at high pressure—for an objective which was never realised. It was designed to detect neutrons from the thermonuclear device ZETA, which had been thought, in the late 1950s, to have made a dramatic break-through by achieving the conditions for thermonuclear reactions. But then a definitive experiment by Basil Rose and two of his colleagues demonstrated that the reactions observed in ZETA were not thermonuclear. In fact this goal, pointing towards the potential source of power for the next century, has not yet been reached. So there was David's beautiful instrument resting in its mount, unwanted. Having been in cosmic ray physics, he had not thought about applications of his detector in the nuclear field, but I saw that, with the availability of high-energy proton beams from the tandem, it could be used to measure neutron spectra from some interesting reactions not previously studied. I suggested that we collaborate on such an experiment.

With some hesitation David agreed, and set up his big ionization chamber in the tandem target room. It was extremely heavy, but David, though not a big man, was remarkably strong. He seemed to be the only person able to lift his monster. Working with David I found quite exacting: he was a cautious and meticulous experimenter. However, I was to profit by his example when, later, my work called for some very precise measurements. The experiment with David proved a great success.

At this juncture Eric Paul showed me a letter that he had received from Dr Muirhead, a theoretical physicist at Liverpool University. It mentioned briefly an important theory which had recently been put forward by two prominent

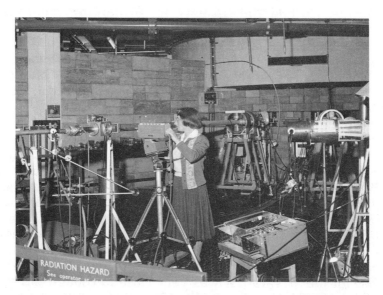

Figure 12.2 The author working in one of the target rooms of the tandem accelerator (1961). (Courtesy of Harwell Laboratory)

American physicists, Richard Feynman and Murray Gell-Mann. It concerned the process of nuclear beta decay, and Muirhead pointed out that some experimental tests of the theory would be highly desirable. 'This is something you might be able to do,' Eric remarked.

The relevant experiments required the accurate measurement of the total energies and half-lives in beta decays of a type known as superallowed Fermi decay. One particular case, that of the oxygen isotope of mass 14, had already been studied (by Willie Fowler and Charlie Barnes) at Cal Tech. But more cases were necessary. One of these was aluminium-26; I could see that this might be done by bombarding a target of magnesium-26 with protons from the tandem so as to form aluminium-26 by neutron emission, the neutrons being measured with David's detector.

But what was it all about? I tried to understand the highly theoretical paper of Feynman and Gell-Mann, but could make little of it. No one at Harwell was sufficiently versed in the fundamentals of beta-decay theory to be able to help me. But, I reflected, I knew someone who was: my mind flashed back to my pleasurable encounter with Roger Blin-Stoyle while he was on sabbatical at MIT. So I went to see him in Oxford, discovering afresh what an attractive and approachable person he was. He had the capacity for explaining complicated theories in simple terms, and although I still did not understand the details, he conveyed to me a feeling for the fundamental importance of Feynman and Gell-Mann's theory. If I could obtain an accurate result for the *ft* value of aluminium-26 decay (*f* being a function calculated from the energy release in the beta decay, and *t* the half-life) then this would constitute a direct test of the theory, Roger said.

I persuaded David to collaborate with me on the energy measurement, while the setting up for the half-life determination was undertaken by the versatile John Montague, with Bob White, the New Zealander whom I had met in Buechner's Lab at MIT and who was now a Harwell Research Fellow. It was a challenging experiment, with many pitfalls to be avoided and corrections to be allowed for, but at last we had a value for the energy release, from which we calculated the *f* function, and a result for *t*. Excitedly we multiplied the two together for our *ft* value. It was very close to though not quite in agreement with the oxygen-14 value.

Roger Blin-Stoyle greeted our result with great enthusiasm, emphasising its significance. 'Could you make measurements for some other cases?' he asked. 'That would be really worthwhile.' He talked about the growing interest in the weak interaction—the force responsible for nuclear beta decay and the decay of elementary particles—and spoke of its recognition as one of the four basic forces of nature, along with the strong interaction which bound the particles in the nucleus, and the more familiar gravitational and electromagnetic interactions. While we proceeded with further experimental work on the superallowed Fermi decays, he would like to study closely the relevant theories, Roger said.

So began an extremely fruitful collaboration. Roger understood about as little of experimental physics as I did about the theory: once I asked him if he would like to look at our tandem accelerator. 'Not really, thanks,' he replied, apologetically. 'If you've seen one, you've seen them all.' But as a closely knit, complementary team, we were enormously successful. Clearsightedly Roger interpreted the experimental data in terms of the continuing theoretical developments, injecting a number of his original ideas. I could follow many of his patient explanations, though there were some aspects for which I just had to take his word. In my unfolding story I will have to ask the reader to do likewise.

Towards the end of 1962 Roger was appointed Professor of Theoretical Physics at the University of Sussex, from where he made frequent visits to Harwell to consult with our beta-decay group. At about the same time David West left the group to work in high-energy physics, and John Montague was required for other jobs. But we gained an important member in Bill Burcham, who, after spending part of a sabbatical year with us, continued to come regularly from Birmingham to join in the experimental runs for the duration of the group's existence. With characteristic modesty he used to joke about being supervised by his former student, but in fact his wisdom and experience were extremely valuable to us. Moreover, his participation was also to prove vital to the survival of the group, as I will presently relate.

We made measurements on a total of ten different beta decays, improving our techniques as more and more accuracy was required by the developing theories. The work was quite fascinating: each case was different, needing special treatment and stretching our ingenuity. While Burcham and I provided the continuity, our group was strengthened and enlivened by a succession of temporary members, mostly Harwell Research Fellows on 3-year appointments. Each new arrival brought with him fresh enthusiasm and ideas. Bob White, for instance, suggested that, instead of David's ionization chamber, we could use the tandem's new magnetic spectrograph; the technique we developed in this way proved highly satisfactory. Another New Zealander, David Robinson, introduced us to the use of computers. David also played the bassoon with great competence; the unusualness of his instrument was surpassed only by Don Gemmel, an Australian Fellow, who worked with Arnold Jones and once startled the lab with a performance on the didgeridoo (a very long Aboriginal horn) that he had brought with him.

Being still an Aussie at heart, I much enjoyed having Australian Fellows in our group. One of these, Greg Clark, a young man of considerable charm, belied his somewhat dilettante manner with some beautiful work on the beta decay of oxygen-14. Another was John Jenkin, through whose initiative and drive we acquired from the USA an important, and in the 1960s very new, kind of gamma-ray detector. Its transport was a tricky exercise since it would have been ruined if not kept constantly at liquid nitrogen temperature. It was sent in a large dewar, and John was worried that the liquid nitrogen might all evaporate while it was in transit. He received a frantic phone call from a customs officer at London Airport saying that it had arrived with smoke coming out of it. To John this meant that at least there was still liquid nitrogen in the vessel, forming a cloud of frozen water vapour when it was shaken up. Fearing now that they might point a fire extinguisher at it, John tore down to the airport in his car to collect it. He said afterwards that he had never seen anything pass through the customs formalities so quickly.

Another asset to our group was my acquisition successively of two Oxford D Phil students for whom I was able to act as official Supervisor when Oxford finally decided to allow Harwell staff to fulfil this role. Having missed the opportunity of formal teaching, I found this an agreeable compensation, and enjoyed seeing both young men develop into successful research physicists.

Meantime, policy changes at Harwell started to produce some serious distractions for me. It all began in 1966 when Walter Marshall, by then Deputy Director, called a meeting of all senior staff to announce that the Establishment, now having diminished responsibility for reactor research, must diversify its interests and seek some financial returns for its expertise. We must sit on a three-legged stool, supported by nuclear work, activities sponsored by Government Departments, and non-nuclear work paid for by British Industry. Fundamental research, to be renamed underlying research, should be directed into fields likely to reap ultimate financial rewards. 'We must be mercenary, not missionary,' Marshall said. 1966 was altogether a year of change as far as I was concerned. Dr Bretscher, whom I greatly respected, as a scientist and as a personality, retired. Basil Rose succeeded him as Division Head, Eric Paul having left earlier for a Chair in Manchester. In that same year Arnold Jones went to Oxford University; I missed his wit and wisdom.

For an Establishment as large as Harwell, altering course was a slow, laborious business, approached with reluctance by many of the dedicated scientists. Basil, being hard pressed by the Directorate, began to look askance at our beta-decay work: though he appreciated its basic scientific merit, he realised that it did not fit the new definition of underlying research. However, we were reprieved when Harwell decided to charge Universities for their use of its facilities. The Science Research Council (SRC) agreed to pay for the considerable number of University users of our tandem; thus Burcham was able to bring with him financial support for his participation in the beta-decay work, ensuring the group's survival as well as continuing to make valuable contributions.

My next hazard involved a serious threat to the tandem itself. We had already survived a contest with Aldermaston over which of the two Authority tandems should be shut down; they lost. Then, in 1970, the Harwell Nuclear Committee, being required to make some economies, decided that one of the four major accelerators should be closed down. The variable-energy cyclotron (VEC), which belonged to Chemistry Division and was Harwell's most recently built accelerator, justified its continuing existence on the grounds of its extensive use by members of the Metallurgy Division for simulating neutron damage of structural materials in the fast reactor; the linear accelerator was doing reactor-oriented low-energy neutron work; and there were plans to modify the high-energy cyclotron (by then referred to as the synchrocyclotron) to produce a fast-neutron source for reactor research. So the Committee decided that the tandem, although it was bringing in a substantial income from its external users, should be axed.

It was a remarkable train of circumstances that saved us.

Just at this juncture the metallurgists, who had been using the proton beam of the VEC to study irradiation damage in various steel alloys, decided that what they really needed for this purpose was an accelerated beam of iron ions. The VEC engineers said it couldn't be done without considerable development work, which might take several years. Hearing of this, Humphries and I consulted Ralph Dawton, who, after having completed his work for the Oxford project with Doug Allen, had returned to us to develop the tandem ion sources. Could he make us a source of iron ions, we asked. Ralph loved a challenge of this sort and said he'd see if he could modify one of his present bench models. In an incredibly short space of time Ralph had a prototype source of negative iron ions running in his experimental lab. It was a splendid achievement. The metallurgists were impressed, and financed our efforts in installing the source on the accelerator and studying the accelerated beam characteristics; Brian Hooton, one of our able young physicists, helped me in these studies. Soon the metallurgists were able to do some successful irradiations, and the tandem's continued existence was assured.

The VEC Group, now seeing themselves as a possible victim of the economy drive, decided that they might after all be able to make an ion source and injector for their machine which could satisfy the metallurgists. At the same time Basil Rose, who had been seeking to protect both the synchrocyclotron and tandem from the threat of closure, introduced a scheme of joint operation, with alternate running of the two machines by shared operators. This imposed a severe strain on our accelerator resources, what with the reduced total running hours plus a large period of time to be devoted to the metallurgists. In 1974 the system of joint operation was abandoned, and a few years later the synchrocyclotron finally succumbed to closure. Meantime the metallurgists decided that the VEC, having by then produced the required ion beam, was more suitable to their purposes. The tandem reverted to its normal regime, now safe in its newly established reputation for value and versatility, and its beleaguered staff heaved a sigh of relief.

During this traumatic period I was buoyed up by the increasing excitement of our beta-decay work. Other laboratories were being attracted into the field; Denys Wilkinson and his colleagues in particular made substantial contributions. At the same time a dramatic new theory emerged. It was the first step towards the goal which is still very much occupying the minds of theoretical physicists: the unification of the four forces of nature—the weak and strong nuclear forces, and the electromagnetic and gravitational forces. In this first step a unified weak and electromagnetic (electroweak) interaction was proposed by the theorists Weinberg and Salam. Based on a so-called quark model for the constituents of protons and neutrons, their theory predicted the existence of heavy elementary particles called the W and Z bosons which propagated the weak interaction in the same way that the photon acted as carrier for the electromagnetic interaction.

In 1970 Roger Blin-Stoyle cleverly perceived that a link could be established between our beta-decay data and a theory of this kind, via the electromagnetic effects known to be associated with the beta decays. Working together, we inserted my experimental results into his calculations and were able to publish an important paper showing that the beta-decay interaction could only be explained by a theory which included a heavy W boson. Subsequently an American theoretician called Sirlin made a calculation of the electromagnetic effects specifically in terms of the Weinberg–Salam electroweak theory. Then, in 1975, Wilkinson, following Blin-Stoyle's lead, drew together the latest superallowed Fermi decay data, some recent elementary particle decay data, and Sirlin's calculations, to show that the experimental results were consistent with a specific quark model and with a Z boson of the predicted mass.

From beta decay to unified forces and W and Z bosons! This was heady stuff for a simple experimentalist like myself. I counted myself very lucky to have got involved in what proved to be such exciting and fundamental work. It was in fact to take another eight years for the W and Z bosons to be observed directly, with the very high energy accelerator at CERN: a Nobel Prize-winning effort, as was that of the protagonists of the electroweak theory.

Then came an event which was the biggest surprise of my career. A letter arrived for me from the President of the British Institute of Physics telling me that Roger Blin-Stoyle and I were to be awarded jointly the Rutherford Medal for 1976. I couldn't believe it. This prestigious award was bestowed every two years for contributions in nuclear or elementary particle physics. No other woman had won it, and only

Figure 12.3 Professor Roger Blin-Stoyle and the author receiving the 1976 Rutherford Medal award from Sir Brian Pippard, President of the Institute of Physics. (Courtesy of Tomas Jaski Ltd)

one other Australian—Sir Mark Oliphant; Denys Wilkinson had been selected in 1962, deservedly so for all his achievements in nuclear physics. That Roger merited such recognition I had no doubt, but I did not feel worthy of it myself. However Roger, delighted at the news, insisted, as he always did, that he couldn't have made progress without my work; it was the unusually close collaboration between theorist and experimentalist that had won the day for us. I looked for hints of "why her?" from my colleagues, but found them all generous in their approval. Basil Rose and Denys Wilkinson seemed particularly pleased, and I suspected them of having had a part in the original nomination. So I abandoned myself to my sense of exhilaration, wishing only that my mother, who had died three years previously, could have witnessed this highlight of my career.

Chapter 13

Changing Times

The limelight in which our beta-decay group suddenly found itself, as a result of the Rutherford Medal award, was somewhat embarrassing. We had been carrying on unobtrusively—maintaining a low profile, as a currently popular cliché had it—because, although Burcham's SRC support gave us some justification for continued existence, our work was clearly at variance with the Harwell criteria for acceptable underlying research. 'I really think you should close the group down,' said Basil Rose anxiously. As it happened, this was a natural moment at which to do so. We had achieved as much as we could with our existing apparatus. To make significant further advances in the superallowed Fermi decay studies, we would have had to make even more accurate measurements, requiring the building of a new, more sophisticated experimental system; I was not enthusiastic about a long-term enterprise of that kind. At the same time Burcham felt that he should not continue to ask the SRC for funds. So, in 1977, the beta-decay group came to an honourable close.

I still had some interesting experimental work in hand, in another field more closely allied to Harwell's interests. This was the study of atomic interactions of fast heavy ions in materials. It began in 1970 when Brian Hooton joined me in analysing the components of the accelerated iron ion beam that we produced from the tandem for the benefit of the metallurgists. Brian pointed out that there were some interesting experiments to be done on the behaviour of fast heavy ions in passing through thin foils, and that such studies might well have some practical applications for the future. So, in parallel with the beta-decay work, I pursued these studies with Brian for several years. I enjoyed the work, and felt at the same time that this more acceptable kind of underlying research compensated to some extent for my indulgence in my beta-decay experiments.

In 1974 Brian was asked to transfer for a two-year spell to the Commercial Office, which had been set up to promote money-earning activities in Harwell. However, I was not left without a collaborator since, just at that time, the Tandem Group acquired a new Harwell Research Fellow, Carl Sofield, from Canberra. Carl was the last of these Research Fellows and a worthy representative of the long line of Ph D graduates that we had had over the years from the ANU, Canberra; they all bore the hallmarks of practical capability and a spirit of "get-up-and-go".

Carl had been working in atomic physics on a subject called beam-foil spectroscopy, so it was appropriate for him to join me in carrying on where Brian Hooton had left off; he threw himself enthusiastically into studies of the processes by which heavy ion beams from the tandem lost energy in passing

Figure 13.1 Support staff of the tandem accelerator with a few of the research scientists (1976). Front row, second from the right: Peter Humphries (Chief Engineer); third from right: Carl Sofield; fourth from right: Eddie Mould (Senior Technician). Back row on extreme left: Ralph Dawton; immediately right of Carl: Nicholas Cowern. (Courtesy of Harwell Laboratory)

through thin foils. It gave promise of being an interesting topic to pursue, so, towards the end of 1975, I decided to take on another Oxford D Phil student, Nicholas Cowern, to work with us on this project. Nick had a sound knowledge of theoretical atomic physics and was able to make some interesting original contributions to the interpretation of the experimental data. I had to work hard to keep up with Carl and Nick in a field which had been unfamiliar to me; but my efforts were to prove well worthwhile.

By the time 1977 arrived I had to face up to a fact that I had been pushing to the back of my mind. It was a question of what I called "technical" sex discrimination, which seemed to be confronting me seriously for the first time in my career. My sixtieth birthday was due in January, 1978, and, although my contemporaries who had joined Harwell at the same time as I did had been appointed to age 65, my contract, because of my sex, had specified the age 60 for its termination. I took little notice of this at the time, not imagining that I would still be at Harwell some 27 years later. But now, I thought, in view of all the recent legislation establishing equality of opportunity for the sexes, the UKAEA will surely extend my appointment for five years, to put me on a par with my male colleagues. The official response came as a shock: they had redressed the inequality by making all recent contracts for Harwell employment terminate at age 60, irrespective of sex. There was no question of my contract being extended, they said.

Seething with indignation, I made representations to the Harwell Directorate about it. They said that they appreciated the injustice, but could do nothing about extending my appointment. Instead, they offered me a special Consultancy, renewable for five years from the official termination of my employment. Although I would have to relinquish my position as Group Leader for the Tandem Group, I would be free to come and go as I pleased, and to carry on with research work as well as acting in a consultative capacity. The Consultancy fee would not be large, but I would concurrently be drawing my superannuation.

So what was I to do? Should I chain myself to the Main Gate and fight for my right to continuing full employment, or should I accede to Harwell's proposition, accepting the opportunities afforded by the Consultancy and thereby maintaining good relations with AERE? Not having strong feminist tendencies, and feeling that I would probably be fighting a losing battle, I did not seriously consider the former alternative. Thus on the last day of January, 1978, I handed in my pass and took formal leave of Harwell with a heart-warming retirement ceremony organized by the Nuclear Physics Division. The next day I returned, with a new pass, for a run on the tandem with Carl and Nick. During the following year a British woman, whose profession I cannot now recall, took her case, which was similar to mine, to the European Court of Justice, and finally won it. But I couldn't believe that it was worth the long struggle that she had had to go through.

My successor as Tandem Group Leader was Tom Conlon, an experimental nuclear physicist who had joined the Tandem Group as a Harwell Fellow in 1966. He had a pleasant, equable temperament and readily accepted my continued presence in the building. I soon discovered the sense of relief in being free of bothers about budgets, staff problems, progress reports, meetings, and endless paper-work. I was able to give my full attention to the heavy ion atomic physics work, and to see it develop and prosper as more members joined the group. Altogether, 1978 was a satisfying year for me, with continuing participation in physics research, the ability to visit University Physics Departments where I was invited to give lectures, and, at the same time, the freedom to indulge in outside activities, like water-colour painting, which I had long been wanting to pursue.

Meantime, John was still working full time—his appointment being to age 65—but with increasing frustration as his interests diverged from Harwell policy. He had taken on the job of designing and building an autoguider for optical telescopes, and had installed one at Cambridge. He had prospects of selling others also, but, seeing how happy and carefree I was with my new status, he decided to take early retirement from Harwell, in March 1979, and to accept a part-time Consultancy at the Royal Greenwich Observatory, with work in a pure research atmosphere more to his liking. Having been a rebel as far as administrative formalities were concerned, he had some difficulty with his official severance from Harwell. This had to do with the instruments that he had acquired over the years, particularly for use in joint experiments he organized with groups at Jodrell Bank, Cambridge, Dublin, and as far away as Bologna in Italy. The list of these instruments recorded by the Harwell Inventories Section showed discrepancies, in serial numbers and the like, with the items John finally returned to them. In fact, to their dismay, he returned more items than were on their list. His final month's salary was withheld for some time while they tried to sort out the muddle to their satisfaction. So John sent the Pay Office a note requesting that the outstanding payment be placed meanwhile in an interest-bearing account. This apparently was too much for the system to cope with: they promptly paid him his money.

We thoroughly enjoyed those Consultancy years, and the opportunities to make extended trips together to Australia, America, South Africa and Europe, visiting Laboratories, giving lectures, and making contacts with numerous old friends and associates. Especially pleasurable were reunions organized for me at my old school, at Sydney University, at the Radiophysics Laboratory, at MIT, and in the homes of many of the Fellows, visiting scientists, and former staff members that I had worked with at Harwell.

In the atomic physics work, I saw Nick Cowern successfully complete his D Phil thesis, and Carl, having been appointed to a permanent staff position, begin to exploit the heavy ion techniques not only in basic research but also in applications to commercially viable projects. The scattering of these ions from a solid surface proved to be an effective means of analysing the nature and distribution of small amounts of contaminants or additives in the specimen. The method continues to be used to good and profitable effect. Other ways of applying the tandem to commercial projects were devised by Tom Conlon and his colleagues. The now venerable machine (the longest running tandem accelerator in the world) is still being successfully used some thirty years after it first came into operation. I feel glad that it has adapted well to the changing face of Harwell, but thankful that I was able to enjoy it in its hey-day of fundamental research.

Even when our Consultancies formally came to an end, John and I were able to maintain our contacts with Harwell by undertaking work for an internationally based system of computerized indexing of scientific journals. We continue to go into Harwell once or twice a week for this purpose and I keep in touch with the activities of the Tandem Group.

As far as hobbies in semi-retirement are concerned, one which has proved very successful was unexpectedly thrust upon me. When John decided to take early retirement from Harwell, I said that he should find himself some interesting activity apart from physics. I thought he might perhaps take up golf, or bowls, or some similar pastime. To my surprise, the house started to fill up with magazines on sailing, and a latent interest, of which I hadn't been aware in John, came to the fore. He persuaded me to go on a sailing course with him, and then, in 1982, he acquired a 26 foot yacht which was berthed at a marina on the Hamble River, close to the Solent. He couldn't sail the boat on his own, so, somewhat apprehensively, I agreed to cooperate. Thus, as mate, deck-hand, and galley-chief, I began to learn with him the art of sailing. I have found it a challenge; but, while drawing the line at going out in strong winds, I share with John the continuing interest and satisfaction to be derived from it, and, after an invigorating day's sail, I find there's nothing quite like dropping anchor in a quiet, unspoilt spot, such as Newtown Creek, on the Isle of Wight, and listening to the chattering of the gulls and waders as the sun sinks on a balmy summer's evening.

My other absorbing activity has been writing: papers, historical articles about Harwell, and now, my own story. In surveying the course of my career, I have realised my great good fortune in having been able to fulfil so completely my childhood ambitions. I have enjoyed my professional life enormously: in particular, working in a laboratory; interacting with similarly motivated colleagues; and, above all, gaining, through some knowledge of physics, extra insight into the beauty and mystery of the physical world, and into the remarkable achievements of brilliant scientists.

Figure 13.2 John Jelley and the author on a sailing course in 1980.

Women in Physics

Many well-informed articles and books have been written under this general title by people who have carried out statistical studies and extensive research on various aspects of the subject: for example, reasons for the paucity of women physicists: problems facing women who enter the profession; and the advantages they can expect from learning physics and following it as a career. I could not presume to produce an authoritative survey of these topics: I am no sociologist, psychologist, educationalist, or historian. Is there, then, anything new that I could contribute to discussions on the subject of women in physics? Well, having been one of the few—immersed in physics throughout my professional life—I can at least approach the topic from the inside. Thus, even though likely to be biased, the views I have formed from my own experiences, and from those of other women physicists I have known, may be of some interest. So here are my thoughts about sex discrimination; about why girls tend not to pursue physics at school; and about the rewards I can see in studying physics.

In *Physics Today*, the monthly journal of the American Physical Society, many advertisements appear for posts in American University Physics Departments. They usually end with a sentence such as 'We especially encourage applications from women, members of ethnic minority groups, and disabled persons.' My hackles tend to rise at the sight of such a statement, which indicates that the University is in need of filling up its specified quota of individuals in these apparently underprivileged categories. What an indignity! How does a self-respecting woman physicist feel if she is appointed? Her male colleagues look askance at her, thinking that she was selected in order to meet the quota, rather than because she was necessarily the best person for the job. One woman I know told me how difficult it was for her to establish herself on her merits in her new job. There have even been instances where a male applicant, complaining of what is called "reverse discrimination", has brought a case against the University on the grounds that his qualifications were superior to those of the woman who was appointed.

Yet the quota rule was a hard-won concession achieved through a vociferous campaign by feminist movements and supporters of minority rights. There seems little doubt that sex discrimination has been a genuine handicap to some American women physicists, and it is said still to exist, particularly in cases of promotion to tenured and more senior positions. However, I have not heard personally of complaints about this problem from the (admittedly few) women physicists that I have met in America. Janet Guernsey, for instance, who became Head of the Physics Department at Wellesley College as well as playing a prominent role in the American Association of Physics Teachers, says that she now sees zero chauvinism and complete respect for female abilities. She talks with pride and satisfaction of the successes of former physics students from Wellesley.

What of women physicists in Australia? Before World War II there were hardly any. The first to graduate in physics from Sydney University, Edna Sayce (B Sc 1917) had considerable difficulties, as her daughter, Margaret Wright, also a physicist, has related to me. At the end of her first year she decided that she liked physics so much that she wanted to go on with it. Professor Pollock, the Head of Department, was implacably opposed to the principle of women pursuing physics. But she was a strong-minded person, not prepared to take "no" for an answer. 'What was your first year exam result?' asked Pollock. 'I came first,' she replied, at the same time producing a sample of her carpentry, a skill learned from her carpenter father. The professor grudgingly admitted that he couldn't stop her from entering second year. After graduation she was appointed Demonstrator in physics—Pollock was away on active service at the time, and Acting Professor Vonwiller was quite happy to take her on. Later she married Dr George Briggs, the lecturer who inspired me about nuclear physics.

The next woman at Sydney University to graduate in physics was Phyllis Nicol, the story of whose tribulations I have already related. I think her problem was not simply a case of sex discrimination but also a shortage of jobs. By the time Ruby Payne-Scott and I came along, World War II had completely transformed the situation. Many jobs were opened up for physicists, and neither of us observed significant signs of personal sex discrimination from our colleagues at the Radiophysics Lab. However, technical regulations about the employment of married women certainly affected Ruby; and Rachel Makinson, whose career I mentioned briefly in Chapter 4, suffered even more serious consquences from them. Moreover, although she was accepted on equal terms by her working colleagues, and her considerable scientific achievements were fully respected, Rachel came up against blatant male chauvinism on the part of a particular Section Head, and feels that her ultimate advancement to Chief Scientist and Acting Division Chief would have come sooner and more readily had she been a man.

Recent Australian legislation has eliminated the old discriminatory regulations, establishing, in principle at least, equal employment opportunities for women; also, sociological attitudes are changing steadily, and I believe that Australian women ambitious to embark on a career in physics now have a much easier path to tread than used to be the case.

When I came to England in 1946, I gained the impression that sex discrimination in general was not as serious a problem as in Australia. Particularly after World War II, during which women played such a vital role in the Armed Services, Industry, Public Service, and the professions, their place in society seemed to have been revolutionized. I noticed that at social gatherings there was far more mixing of the sexes than was usual in Australia. Even in Cambridge, that bastion of male superiority, I witnessed in 1948 the admission of women to full membership of the University with hardly a murmur of dissent. In the Cavendish, I found no sign of sex discrimination. For a woman like the remarkable scientist, Dorothy Hodgkin, there was universal respect. Another outstanding woman, the astrophysicist, Jocelyn Bell, who, as a Cambridge research student became famous in 1967 for her radioastronomy observations which led to the discovery of pulsars, was acclaimed and admired. She is now in a senior management position at the Royal Observatory Edinburgh.

At Harwell I was fully accepted as a physicist. There was what I call "technical" discrimination, in that to start with I was paid less than my male counterparts; but that disparity was removed before long, and subsequent legislation has provided for equality of opportunity for the sexes. This is not to say that the effects of male chauvinism do not still exist. Some of the young women physicists that I have recently met at Harwell have told me that they encounter signs of it in subtle ways; but they are not aware of sex discrimination seriously affecting their jobs and prospects. It seems that nowadays British women face no greater discriminatory problems in physics than they do in other male-dominated professions. Ironically, a young woman training as a house doctor in a prominent Oxford hospital told me recently that she had no difficulties with her male colleagues, but met hostility from some of the nurses who, used to their traditional relationships with the male doctors, resented taking orders from a woman.

Statistical surveys indicate that the percentage of women physicists in higher administrative and top management positions is significantly less than that of men. Sex discrimination may account partly for this situation, but there are other factors also to be considered. Married women in particular are frequently hindered by the necessity of changing jobs because of their husbands' movements, and by family responsibilities. In some cases they may not be ambitious to attain top management levels. I certainly had no desire for this; my observations suggested that men promoted to the upper grades often seemed harassed and less happy than I was. I admit to having enjoyed my position at Harwell as a Group Leader, a function which I feel I performed successfully; but it was of overriding importance to me that the Group was not too large for me to be able to pursue some basic research work as well.

It is of course just as well that not all women physicists share my attitude. Men and women with vision and courage in influential positions are vital to the promotion of scientific research and its applications for good in the modern world; also to the furthering of education in science. More leaders are needed, and it is important to have gifted women as well as men in these positions. I particularly admire the achievements of the few such women that I have known: for example, Janet Guernsey in America, Rachel Makinson in Australia, and Daphne Jackson in Britain.

Professor Daphne Jackson stands out for me as a successful and important woman physicist. An unassuming but quietly determined person, she has not only made a variety of contributions to physics research but is now Head of the Physics Department of Surrey University and was for a time Dean of the Faculty of Science. She is an effective champion of

the cause of women in physics; for example, during the last
few years she has initiated and coordinated a successful
Fellowship scheme for helping qualified women to retrain and
then take up professional jobs in physics or engineering after
a career break. The scheme is funded to a large extent by
industry, which is conscious of the present shortage of competent physicists required to meet a growing demand in industry
and research. Other organizations such as the Women in
Physics Subcommittee of the Institute of Physics are active in
helping professional women physicists and encouraging more
schoolgirls to take up physics. Similar bodies in America and
Australia are working to these ends. It is interesting to note
also the male support now coming from University Physics
Departments. Conscious of the drop in student/staff ratios
(mainly as a result of a fall in birthrate), they are keen to
have girls helping to make good this deficiency.

This brings me to the question of why so few girls do take
up physics. Many reasons have been offered by way of explanation, including innate lack of aptitude in girls compared
with boys; parental influences and social pressures encouraging girls to develop "feminine" interests and characteristics;
inadequate teaching; and the prejudice of boys. Looking at all
the arguments, one might conclude that girls who do become
physicists must be biological deviants, uncommunicative, unsociable, inhuman, unfeminine, overbearing, and well prepared to do battle against the barriers of male chauvinism.

The women physicists I have known certainly do not fit
such a picture. Their personalities are quite diverse; their
most notable common characteristics are perseverance, self-
reliance, and a positive enthusiasm for their subject. In most
cases they have received early encouragement from a parent,
relative, or teacher. Nevertheless they have inevitably met
some deterring influences along the way, so it is not surprising that, having surmounted such obstacles, they have proved
to be particularly able physicists. This fact now seems to be
generally recognized. In recruitment at Harwell, for instance,
some of my male colleagues have remarked that they are
specially interested in female applicants, anticipating that
they are likely to be capable and well motivated. Women
physicists now employed at Harwell seem to bear out these
expectations. There are still not many of them, though the
numbers have grown in the last decade or so; I know three in
Harwell's Nuclear Physics Division, representing the younger
generation. One of them, Ruth Harper, who completed her
post-graduate research in solid-state physics at the University

of St Andrews in 1982, is now manager of a small section concerned with materials science research and the development of special detectors particularly for nuclear applications. She clearly has the attributes which I have noted above; these qualities are likewise evident in two younger women in the Tandem Accelerator Group.

As far as the question of girls' aptitude for the physical sciences is concerned, I do not believe that biological differences between girls and boys can be important because in some countries the situation is quite different. In Poland, for instance, several physicists have told me that there are as many women as men in the field. British girls are said to show, on average, less scientific ability than boys. But I wonder if this difference is significant compared with the differences between individuals of either sex. When, long ago, I offered my finally discarded Meccano set to the young brother of one of my school friends, I was disgusted by his lack of interest. Even among physicists there are vast differences in types of ability: compare, for instance, a man like the theoretical physicist, Pauli, of whom it is said that he had only to walk past a lab where there was some delicate apparatus and it would break, with Rutherford, who was highly suspicious of theoreticians and relied on the results of his experimental observations.

I believe that the most important factors deterring girls from taking up physics are the ingrained social traditions, and the results of these in the education system. In Victorian times it was "nice" for girls to study flowers and butterflies— hence the popularity of the biological sciences with women— but not nice for them to concern or dirty themselves with mechanical objects, which boys still tend to adopt as their prerogative. Traditional attitudes are mostly illogical but die hard. For example, cookery is regarded as a girls' subject at school; yet the Chef in a high-class restaurant is almost invariably male. Instinctively society in general, and parents in particular, tend to treat girls and boys differently, from infancy, and to encourage different characteristics to emerge. Unequal opportunities for girls to learn physics at school have likewise followed from these traditions. But the situation seems to be improving. Enlightened parents and teachers are helping to further the participation of girls in the physical sciences.

I turn now to why I think that girls should be encouraged to study physics.

In the first place, in my own experience, and that of other women physicists I have known, a career in physics is absorbing and fulfilling, whether it be in research, technological developments, or applications in the many fields like medicine, or the environment. A vast range of jobs is on offer to people with scientific qualifications. The option to pursue such a career should be kept open for girls as well as boys through some early teaching of physics at school; those who develop a natural enthusiasm for the subject should be encouraged to continue.

Even for girls not contemplating a career in physics, some basic knowledge of the subject is an advantage in this scientifically oriented age, helping them to understand the functioning of the many mechanical, electrical, and electronic devices that they nowadays encounter; also so that they can take an informed interest in debates about topics such as the environment, pollution, and energy sources. The voice of women able to take an understanding and balanced view of present-day problems in these areas should be heard.

Finally, I would like to reiterate my theme of previous chapters about the new perceptions that some knowledge of physics gives of the fascinating nature of the physical world. A rainbow, for example, is a beautiful phenomenon. If you can understand how it is formed by the refraction of the sun's rays in water droplets, then another dimension is added to your enjoyment of the effect. At a more profound level, I am reminded of the American lady whom I met in the States in 1960. She remarked that she did not like astronomers because they destroyed her childhood illusion of stars being holes in the floor of heaven, through which shone the brilliant light beyond. But how much more imaginative is the picture which modern astronomers have given us of our Galaxy, and the awe-inspiring grandeur of the Universe. In stretching the mind to grasp the significance of discoveries and interpretations in the rich tapestry of physical phenomena there is an unending source of intellectual pleasure.

I once received from Denys Wilkinson (now Sir Denys, recently retired Vice Chancellor of the University of Sussex) a letter, at the foot of which was printed the following quotation:

'. . . we delight in physics . . .' Macbeth II (iii)

I looked up the reference; the Shakespearian text read:

'The labour we delight in physics pain'

Denys' joke serves to underline two very real aspects of my philosophy: physics is a delight, and physics is fun.

Index

224